Waste and Distributive Justice in Asia

T0173945

Conflicts over waste disposal facility siting is a pressing issue not only in developed countries but also in fast-growing countries that face drastic waste increase and rapid urbanisation. How to address distributive justice has been one of the biggest concerns.

This book examines what determines the influence of distributive justice in siting policy. In the 23 wards of Tokyo, one idea of distributive justice, known as "In-Ward Waste Disposal" (IWWD), emerged amid the ongoing garbage crisis in the early 1970s. IWWD was adopted as a significant principle, but its influence waxed and waned over time, until the idea was finally abandoned in 2003.

To unravel causes and mechanisms behind the changing influence of IWWD, this book adopts a framework that considers not only ideational causes, but also the power struggles between rationally calculating actors, as well as the influence of external events and environments. By combining an in-depth case study with an integrative theoretical framework, this book tells a thought-provoking story of the changing influence of IWWD in a deep, comprehensive and consistent way. This book provides significant insights and lessons for both academics and practitioners.

Takashi Nakazawa is Associate Professor of Sociology and Politics at the Faculty of Informatics in Shizuoka University, Hamamatsu, Japan. His recent publications include "Politics of distributive justice in the siting of waste disposal facilities: the case of Tokyo", *Environmental Politics*, 25(3), 2016.

Routledge Studies in Environmental Policy and Practice
Series Editor: Adrian McDonald

Based on the Avebury Studies in Green Research series, this wide-ranging series still covers all aspects of research into environmental change and development. It will now focus primarily on environmental policy, management and implications (such as effects on agriculture, lifestyle, health, etc.), and includes both innovative theoretical research and international practical case studies.

For a full list of titles in this series, please visit www.routledge.com/ Routledge-Studies-in-Environmental-Policy-and-Practice/book-series/ ASHSER1048

Protected Areas and Regional Development in Europe
Towards a New Model for the 21st Century
Edited by Ingo Mose

Energy Access, Poverty, and Development
The Governance of Small-Scale Renewable Energy in Developing Asia
Benjamin K. Sovacool and Ira Martina Drupady

Sustainability and Short-term Policies
Improving Governance in Spatial Policy Interventions
Edited by Stefan Sjöblom, Kjell Andersson, Terry Marsden and Sarah Skerratt

Communities in Transition
Protected Nature and Local People in Eastern and Central Europe
Saska Petrova

Scientists, Experts, and Civic Engagement
Walking a Fine Line
Edited by Amy E. Lesen

Negotiating Water Governance
Why the Politics of Scale Matter
Edited by Emma S. Norman, Christina Cook and Alice Cohen

Storage and Scarcity
New Practices for Food, Energy and Water
Giorgio Osti

Waste and Distributive Justice in Asia
In-Ward Waste Disposal in Tokyo
Takashi Nakazawa

Waste and Distributive Justice in Asia

In-Ward Waste Disposal in Tokyo

Takashi Nakazawa

Routledge
Taylor & Francis Group

LONDON AND NEW YORK

First published 2018
by Routledge

2 Park Square, Milton Park, Abingdon, Oxfordshire OX14 4RN
52 Vanderbilt Avenue, New York, NY 10017

Routledge is an imprint of the Taylor & Francis Group, an informa business

First issued in paperback 2019

British Library Cataloguing-in-Publication Data
A catalogue record for this book is available from the British Library

Library of Congress Cataloguing-in-Publication Data
A catalog record for this book has been requested

ISBN: 978-1-138-57363-5 (hbk)
ISBN: 978-0-367-89151-0 (pbk)

Typeset in Times New Roman
by Deanta Global Publishing Services, Chennai, India

Contents

Figures

Tables

Preface

I began researching conflicts over locally unwanted facilities when I was studying for a master's degree at Hitotsubashi University in Japan. Originally, I was interested in democratic decision making based on my doubts about the legitimacy of majority rule. In majority rule, intensity of preference cannot be adequately reflected in the collective decision. If this is indeed the case, the question becomes, how should we best decide an issue in which a few individuals with a strong intensity preference conflict with the many who have a weaker intensity preference?

I looked into siting conflicts to consider this question in a practical setting, as locally unwanted facilities impose concentrated burdens on neighbouring communities, while their benefits are widely dispersed across society. Siting conflicts interested me as they are deeply related not only to democratic decision making, but also to other significant issues and topics in the social sciences such as community structure, discrimination, social and local activism, public policy making, risk communication, and environmental problems. I have since been studying conflicts over municipal waste disposal facilities, wind turbines, nuclear power stations, nuclear waste repositories, and nursery schools – especially focusing on distributive justice, which I believe is the most crucial aspect of siting conflicts.

This research project was originally conducted as part of my Ph.D. programme at James Cook University in Australia. Japan has gone through thousands of conflicts over waste disposal facilities. In Tokyo, concentration of the waste disposal burden on a single ward caused the "Tokyo Garbage War" in the early 1970s, from which an idea of distributive justice known as "In-Ward Waste Disposal" emerged to address the environmental inequalities. I decided to write an in-depth analysis of the 30-year history of "In-Ward Waste Disposal" in Tokyo, believing it to be crucial to understanding NIMBY politics in Japan and a significant contribution to furthering research on siting conflicts and distributive justice. I hope this book will provide both theoretical and practical insights for scholars as well as practitioners who are facing similar problems throughout the world.

Acknowledgements

I would like to express my sincere gratitude to Prof. Hayden S. Lesbirel of James Cook University for continuously supporting this research project. His guidance helped me throughout the duration of my research and writing. I would like to thank Dr. Mark David Chong and Prof. Glenn Dawes, for their insightful comments and encouragement. My sincere thanks also go to my friends and colleagues, in both Australia and Japan, for their support in my study and life, and those who cooperated for interviews and provided me with materials and insights for the research. Without their precious support, it would not have been possible to write this book. Last, but not the least, I would like to thank my family: my parents, Masao and Akiko Nakazawa, and my grandparents, Tadamitsu and Mieko Ueno, for supporting me spiritually and financially throughout conducting this project and my life in general.

Abbreviations

AWI	All Waste Incineration
IWWD	In-Ward Waste Disposal
LULUs	Locally Unwanted Land Uses
MOH	Ministry of Home Affairs
NIMBY	Not In My Back Yard
OWOI	One Ward One Incinerator
SWM	Sustainable Waste Management
TMG	Tokyo Metropolitan Government

1   Introduction

Distributive justice as an essential issue in siting conflicts

Certain kinds of facilities are often opposed by local residents because of negative side effects, while they are claimed to be necessary for the well-being of the wider public. Such conflicts happen over various types of infrastructure such as hazardous waste disposal facilities, nuclear power stations, wind farms, power lines, highways, dams, and airports (Aldrich, 2008; Gerrard, 1996; Hamersma et al., 2016; Lesbirel, 1998; Munton, 1996; Neukirch, 2016; O'Hare, 1977; Rabe, 1994; Rootes & Leonard, 2009; Wolsink, 2000). They also occur over human service facilities such as public housing, prisons, homeless shelters, and halfway houses (Dear, 1992; Hubbard, 2005; Scally & Tighe, 2015; L. M. Takahashi, 1999; Young, 2012). They are called LULUs (Locally Unwanted Land Uses) and local oppositions against them are often referred to as NIMBY (Not In My Back Yard).

Distributive justice is one of the most crucial aspects of the disputes over LULUs. Distributive justice is concerned with fairness in the allocation of costs and benefits. Locally unwanted facilities impose concentrated burdens on neighbouring communities, while benefits from them are widely dispersed over the society. This inherent imbalance between costs and benefits is one of the reasons why the siting of locally unwanted facilities causes the feeling of unfairness and ignites intense local opposition. In addition, these facilities are not equally distributed among communities. Some communities shoulder a disproportionate number of noxious facilities while others host fewer or none. The studies in environmental justice have pointed out that noxious facilities are concentrated disproportionately in minority communities (Bullard, 1990; United Church of Christ Commission for Racial Justice, 1987). This inequitable distribution of LULUs often gives rise to demands to redress unfairness.

How to address distributive justice, therefore, is one of the biggest concerns for both scholars and practitioners of siting conflicts. To deal with the inherent imbalance and the unfair spatial distribution of LULUs, various ideas and policy approaches have been suggested: compensation (Kunreuther, 1986), risk mitigation (Portney, 1991), controls on cross-border shipment of waste (Rabe, 1994), burden sharing by recycling and reducing waste (Rabe, 1994), and schemes to distribute numerous facilities fairly among various communities (Morell, 1984).

Environmental justice studies also have examined what causes environmental inequality and theorised the principles of environmental justice (Pastor JR., Sadd, & Hipp, 2001; Pellow, 2000; Saha & Mohai, 2005).

In-Ward Waste Disposal and its changing dominance in policy

In-Ward Waste Disposal in the 23 wards of Tokyo

"In-Ward Waste Disposal" (Jikunaishori [自区内処理] or IWWD) is an idea of distributive justice that emerged in the early 1970s in the 23 wards of Tokyo.[1] Waste disposal facilities are typical LULUs, and their siting has been increasingly problematic due to population concentration resulting from population growth and urbanisation (Lang & Xu, 2013; Pellow, 2004; Rootes & Leonard, 2009; Rootes, 2007, 2009a, 2009b; Walsh, Warland, & Smith, 1997). Tokyo has gone through a lot of conflicts over waste disposal facilities and how to redress inter-ward distributive injustice had been one of the most crucial issues. In the 23 wards of Tokyo, Koto ward had been imposed with disproportionate burdens of waste disposal (Figure 1.1). Huge amounts of waste generated in the 23 wards had been dumped in coastal landfills next to Koto ward. The ward had suffered from bad smells, traffic jams, and outbreaks of flies and rats. This distributive injustice brought about the first "Tokyo Garbage War" in the 1970s.

In 1971, Koto ward declared IWWD, calling for a more equitable distribution of the burden. This idea of distributive justice called for self-sufficient waste disposal by each ward, meaning that waste from a ward should be disposed of within the ward. Derived from this basic meaning, it had two different but tightly inter-connected requirements: that waste disposal facilities should be fairly distributed

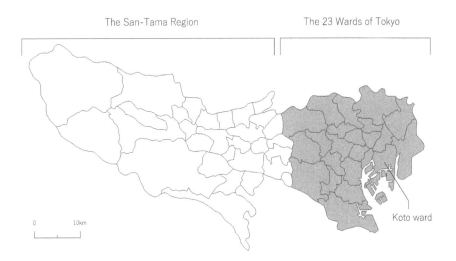

The San-Tama Region   The 23 Wards of Tokyo

0   10km

Koto ward

Figure 1.1   Map of Tokyo Metropolis.

Adapted from a map of Tokyo by CraftMap (URL: http://www.craftmap.box-i.net/).

among the wards and that each ward should be institutionally responsible for the disposal of its own waste.

The first aspect required that waste disposal facilities should be fairly distributed among the 23 wards. As detailed in Chapter 3, IWWD had been exclusively associated with incinerators, because the original claimant of the idea, i.e. Koto ward, regarded insufficient incineration capacity as the fundamental cause of the huge amount of waste dumped in the landfills next to the ward. The aim of IWWD, therefore, was to claim that an incinerator should be sited in every ward so that waste from a ward could be taken care of within the ward. This idea of siting incinerators in every ward was adopted by the Tokyo Metropolitan Government (TMG) and translated into the One Ward One Incinerator (OWOI) policy.

IWWD was not just about a fair distribution of waste disposal facilities, but also about the institutional system of waste management, requiring that each ward should be responsible for and dispose of its own waste. In Japan, each municipality has institutional responsibility for the disposal of waste generated in its own district. In the 23 wards of Tokyo, however, waste management had been administrated regionally by the TMG.[2] IWWD necessitated the shift from this regional disposal to local disposal by each ward[3] by moving the responsibility of waste disposal from the TMG to each ward. Thus, the fair distribution of waste disposal facilities and the self-responsibility of each ward in waste disposal constituted the two components of this idea of distributive justice.

Changing influence of IWWD

However, the influence of this idea of distributive justice on the policies waxed and waned over time until the idea was finally abandoned in 2003. The story of the ebbs and flows is divided into four periods as Table 1.1 shows. The first period is from 1971 to 1973, in which the influence of IWWD rose but only for the distribution of incinerators siting; the influence of the idea of the self-responsibility of each ward in waste disposal was minimal. Koto ward claimed IWWD in 1971, and it was adopted by the TMG. The TMG announced the OWOI scheme and promised to construct 13 new incinerators in the wards with no incinerator or with one but too small to deal with its own waste. On the other hand, the institutional responsibility of each ward in incineration, that is, the devolution of waste management authority and responsibility from the TMG to each ward, was almost ignored in this period.

The dominance of IWWD fell and became weak in the second period from 1974 to 1989. In 1974, the strength of the idea of siting incinerators in every ward started declining despite its strong influence in the first period. The completion of the OWOI policy was postponed in 1974. In 1976, the government relaxed the implementation of this policy and allowed one incinerator to be shared with several wards. Its influence kept declining and IWWD was no longer referred to in waste management policies during the 1980s.

However, the dominance of IWWD revived in the 1990s (the third period from 1990–1996). The TMG showed its renewed determination to site an incinerator in

Table 1.1   Changing dominance of IWWD in policies

Periods	Dominance of IWWD	Distribution of incinerators siting	Self-responsibility of each ward
First period 1971–1973	Strong but limited	• OWOI declared to be achieved by 1975	• The wards' responsibility in incineration was not reflected
Second period 1974–1989	Weak	• The achievement of OWOI was delayed and postponed to the future in 1974 • OWOI was relaxed to joint disposal by several wards in 1976 • IWWD disappeared in the 1980s	• The wards' responsibility in incineration was not reflected
Third period 1990–1996	Strong	• The siting plan based on OWOI was announced	• The responsibility of each ward was adopted and written in the policies on devolution
Fourth period 1997–2003	Weak	• The completion of OWOI was further postponed to the future in the 1997 revision • OWOI was abandoned in 2003	• The devolution of the responsibility was postponed in 1998 • The shift to self-sufficient disposal was given up in 2003

every ward in the 1991 siting plan. Furthermore, an agreement was made among the concerned parties on the devolution of the responsibility in waste disposal from the TMG to each ward. The agreement laid out the roadmap to the self-sufficient incineration system, in which each ward was expected to perform incineration independently. The strength of IWWD came to its climax in this period as both siting incinerators in every ward and the self-responsibility of each ward became influential concepts.

Nonetheless, the strength of IWWD did not last long; it started declining again in the late 1990s and disappeared as the century turned (the fourth period from 1997–2003). The completion of OWOI was postponed to the future in the 1997 plan revision. The devolution of the responsibility for incineration was shelved in 1998. Finally, both requirements were abandoned in 2003. Siting incinerators in every ward was given up, which led to the abandonment of the shift to the self-sufficient incineration system. The influence of IWWD fell despite its prominence in the early 1990s.

Thus, IWWD's influence on policies waxed and waned over time. Why did this particular idea of distributive justice become so influential in siting policy for a few years in the 1970s? Why did its dominance then recede? What made IWWD revive in the 1990s, and why was it abandoned in the early 2000s?

Explaining the changing influence of an idea of distributive justice

This book considers what determines the influence of a particular idea of distributive justice in siting policy and its change over time. While distributive justice is one of the most crucial aspects of LULUs siting, there is no agreed definition of distributive justice to understand and evaluate distributive issues; it can be interpreted in various ways as detailed in the next chapter. Which idea of distributive justice dominates siting policies determines who is to shoulder how much of the burden; different ideas of distributive justice lead to different policies and different outcomes. Political battles, therefore, arise not only over distributive justice in general, but also over which idea of distributive justice to employ. This significance and plurality of distributive justice raises a question: how and why does a particular idea of distributive justice out of many rise to prominence and become influential in siting policy in a particular place and time? While various ideas and approaches have been suggested by scholars and practitioners in siting conflicts and environmental justice, this question has not been fully explored.

IWWD in the 23 wards of Tokyo provides an interesting, thought-provoking case to answer the question. As noted above, the dominance of IWWD in Tokyo fluctuated over time through the four periods. A comparison of the four periods makes it possible to thoroughly examine what caused the diachronic variation in the influence of this idea of distributive justice. Besides, IWWD has taken significant roles in siting conflicts and waste management policy in Japan. This country has seen thousands of conflicts over waste disposal facilities. Cross-border movement of waste is one of the biggest causes of these conflicts (Taguchi, 2003); IWWD is often used as a justification to build facilities in some cases, or as a cause for rejecting siting projects in others. A thorough analysis on the 30-year political history of IWWD in its birthplace, therefore, offers perspectives necessary to better comprehend siting conflicts and waste management policy in Japan.[4]

Siting disputes are becoming urgent issues and attracting much public attention not only in developed countries but also in rapidly developing countries. There are an increasing number of studies of siting conflicts in Japan and other Asian countries (Aldrich, 2008; Fan, 2008; Fung, Lesbirel, & Lam, 2011; Hsu, 2006; Johnson, 2010; Kang & Jang, 2013; Lesbirel, 1998; Munton, 1996; Sun, 2015). However, studies of locally unwanted facility siting are still mostly based on experiences in Western countries, especially in North America. To enrich our understanding on siting conflicts, we need to conduct more studies in non-Western countries and grasp the politics of facility siting in different socioeconomic contexts. As the empirical study of this book shows, the rise and fall of IWWD's prominence in policy was related to various intertwining factors such as economic conditions; policy paradigms in waste management; the political institution of Tokyo; capability of governments; and power struggles between concerned actors such as regional and local governments, labour unions, environmental and local movements, and businesses. An in-depth qualitative analysis and rich descriptions of the study unravel the complicated politics behind the changing influence of

IWWD in Tokyo, and thereby provides significant insights and lessons for both academics and practitioners to better understand waste disposal facility siting and distributive justice policies of Tokyo in particular, and environmental politics of Japan in general.

To explain the changing influence of an idea in policies, this study looks into ideational approaches in political studies. An increasing number of studies incorporate roles that ideas play in policy formulation and implementation into political analysis (Béland & Cox, 2010). However, little is known about what makes an idea prominent in policy making and why a once-dominant idea loses its influence. Furthermore, how an idea's influence is related to non-ideational factors such as interests, power and material conditions has not been fully studied. To disentangle the complicated politics of an idea, this book proposes a new, integrated theoretical framework that takes into account not only ideational causes but also the power struggles between rationally calculating actors, as well as the influence of external events and environments. By combining an in-depth case study with an integrative theoretical framework, this book tells a story of the changing influence of IWWD in a deep, comprehensive and consistent way, and furthers study on the politics of facility siting and ideas in social sciences.

Chapters

The remainder of this book consists of six chapters. The next chapter shows the theoretical framework and methodology of the study. After illustrating multiple ideas of distributive justice in siting, it reviews the literature on the politics of ideas. It is concluded that a comprehensive approach which integrates ideational and non-ideational factors is necessary to explain the rise and fall of the dominance of an idea. Then, four different types of variables, i.e. ideational legitimacy, interests, power of claimants, and exogenous environments, are introduced. The book argues that the dominance of an idea at any point of time is determined by these four variables and the interaction between them. Applying this framework, an in-depth case study, by document analysis and semi-structured interviews, unravels the complicated political processes through which the influence of IWWD waxed and waned. A diachronic comparison of the four delineated time periods makes it possible to better understand what affected the strength of IWWD.

The four chapters that follow empirically examine the rise and fall of the influence of IWWD, period by period. Chapter 3 explains why IWWD became influential in the early 1970s, although it was limited to the idea of siting incinerators in every ward. This idea of distributive justice was advocated by Koto ward, which had suffered the concentrated burdens of waste disposal. The idea of siting incinerators in every ward was quickly accepted by the TMG and translated into the OWOI policy. The rapid economic growth and incinerationism as the long-held policy paradigm provided advantageous environments for the idea to rise to prominence. On the other hand, the self-responsibility of each ward in waste disposal was not reflected in the policies. It was the Tokyo Cleaning Workers Union which prevented this institutional aspect of IWWD from influencing the policies.

Chapter 4 looks into the decline of IWWD in the second period from 1974 to 1989. In spite of its prominence in the early 1970s, its influence started declining from 1974. The TMG was not able to implement the OWOI policy due to the limited land availability in the central part of Tokyo and the persistent local opposition against the incinerator projects. This feedback from the implementation negatively affected the dominance of IWWD. Furthermore, the two oil crises, which occurred in 1973 and 1979, substantially damaged the idea of siting incinerators in every ward. It was not until 1990 that IWWD became influential once again.

Chapter 5 looks at how and why the influence of IWWD revived in the 1990s. Koto ward argued for IWWD once again and launched a campaign to realise this idea of distributive justice. The TMG announced a siting plan to construct incinerators in every ward. The bubble economy in the late 1980s prepared the way for this revival of IWWD. Furthermore, the idea of the self-responsibility of each ward in waste disposal strengthened its influence in this period. The politics of the autonomy system reform of the 23 wards opened up the opportunity for this part of IWWD to influence policies. The dominance of IWWD reached its culmination in the 1990s with the two engines of this idea working together.

However, the dominance did not last long; it started declining from 1997 and IWWD was abandoned in 2003. Chapter 6 considers what caused this decline of IWWD. The economic recession after the financial bubble burst impacted the production of waste and the financial capacity of the governments. The waste management policy changed as Sustainable Waste Management became influential as a new policy paradigm and dioxin problems attracted public attention. Under the rapidly changing circumstances, the idea of distributive justice was losing its influence.

The concluding chapter summarises the empirical results and discusses their implications. It is found that all four variables rose and fell in a synchronised pattern for the idea of siting incinerators in every ward, resulting in very clear changes in the influence of IWWD. The book reveals that this is because economic changes affected the other three variables by impacting both the production of waste and governments' financial capacity. This book argues, however, that this is not economic determinism, and emphasises the importance of considering multiple, different types of variables and examining the interaction between them to explain the prominence of an idea and its change over time.

Notes

1 Tokyo Metropolis is divided into two parts: the special wards and Western Tokyo. The former, the eastern part of Tokyo, consists of the 23 wards. The latter, also known as the San-Tama region, includes 30 local municipalities as of 2013.
2 The institutional responsibility was devolved to each ward in 2000 as explained later.
3 In this book, the phrase "regional disposal" is used to refer to a disposal system in which waste disposal is administered by a regional government (the TMG or a local government association) over the 23 wards as a whole, while "local disposal" means a disposal system in which waste disposal is managed by each ward.

4 IWWD has been understudied despite its importance for siting politics in Japan. Although some studies have looked into the Tokyo Garbage War in the 1970s (Ishii, 2006; Mizoiri, 1988; Osumi, 1972; Shibata, 2001a, 2001b; Shimizu, 1999; Tsugawa, 1993; Yorimoto, 1974, 1977), why IWWD became so influential in this period and why that idea lost its influence quickly has yet to be explained. It is true that there are several studies which analyse IWWD in relation to the regionalisation of disposal, recycling or financial issues (Fujii, 2006; Isono, 2003; Okuda & Thomson, 2007; H. Takahashi, 2001) and these studies are suggestive of what affected the influence of IWWD. However, they lack a comprehensive explanation on how and why the dominance of IWWD in the 23 wards of Tokyo changed over time. This book fills in these gaps in knowledge by conducting an in-depth case study on the changing dominance of IWWD with a consistent, integrative theoretical framework.

References

Aldrich, D. P. (2008). *Site Fights: Divisive Facilities and Civil Society in Japan and the West*. New York: Cornell University Press.

Béland, D., & Cox, R. H. (2010). Introduction: Ideas and Politics. In D. Béland & R. H. Cox (eds.), *Ideas and Politics in Social Science Research*. New York: Oxford University Press.

Bullard, R. D. (1990). *Dumping in Dixie: Race, Class, and Environmental Quality*. Boulder, CO: Westview Press.

Dear, M. (1992). Understanding and Overcoming the NIMBY Syndrome. *Journal of the American Planning Association*, 58(3), 288–300.

Fan, M. F. (2008). Environmental Citizenship and Sustainable Development: The Case of Waste Facility Siting in Taiwan. *Sustainable Development*, 16, 381–389.

Fujii, K. (2006). Seisō Jigyō no To kara Ku eno Ikan: Jikunaishori no Gensoku no Hensen o Tōshite. *Sōkan Shakai Kagaku*, 16, 127–133.

Fung, T., Lesbirel, S. H., & Lam, K. (eds.). (2011). *Facility Siting in the Asia-Pacific: Perspectives on Knowledge Production and Application*. Hong Kong: Chinese University Press.

Gerrard, M. B. (1996). *Whose Backyard, Whose Risk: Fear and Fairness in Toxic and Nuclear Waste Siting*. Cambridge: MIT Press.

Hamersma, M., Heinen, E., Tillema, T., & Arts, J. (2016). Residents' Responses to Proposed Highway Projects: Exploring the Role of Governmental Information Provision. *Transport Policy*, 49, 56–67.

Hsu, S. H. (2006). NIMBY Opposition and Solid Waste Incinerator Siting in Democratizing Taiwan. *The Social Science Journal*, 43(3), 453–459.

Hubbard, P. (2005). "Inappropriate and Incongruous": Opposition to Asylum Centres in the English Countryside. *Journal of Rural Studies*, 21(1), 3–17.

Ishii, A. (2006). Tōkyō Gomi Sensō wa Naze Okottanoka. *Haikibutsu Gakkai Shi*, 17(6), 340–348.

Isono, Y. (2003). Kiso Jichitai to Haikibutsu Shori Hō no Kadai: Jikunaishori o Saikentō Suru. *Gendai Hōgaku*, 5, 47–64.

Johnson, T. (2010). Environmentalism and NIMBYism in China: Promoting a Rules-based Approach to Public Participation. *Environmental Politics*, 19(3), 430–448.

Kang, M., & Jang, J. (2013). NIMBY or NIABY? Who Defines a Policy Problem and Why: Analysis of Framing in Radioactive Waste Disposal Facility Placement in South Korea. *Asia Pacific Viewpoint*, 54(1), 49–60.

Kunreuther, H. (1986). A Sealed-bid Auction Mechanism for Siting Noxious Facilities. *The American Economic Review*, 76(2), 295–299.

Lang, G., & Xu, Y. (2013). Anti-incinerator Campaigns and the Evolution of Protest Politics in China. *Environmental Politics*, 22(5), 832–848.

Lesbirel, S. H. (1998). *NIMBY Politics in Japan: Energy Siting and the Management of Environmental Conflict*. New York: Cornell University Press.

Mizoiri, H. (1988). *Gomi no Hyakunen-shi*. Tokyo: Gakugei Shorin.

Morell, D. (1984). Siting and the Politics of Equity. *Hazardous Waste*, 1(4), 555–571.

Munton, D. (ed.). (1996). *Hazardous Waste Siting and Democratic Choice*. Washington, DC: Georgetown University Press.

Neukirch, M. (2016). Protests against German Electricity Grid Extension as a New Social Movement? A Journey into the Areas of Conflict. *Energy, Sustainability and Society*, 6(1), 4.

O'Hare, M. (1977). "Not on My Block You Don't": Facilities Siting and the Strategic Importance of Compensation. *Public Policy*, 25(4), 407–458.

Okuda, I., & Thomson, V. E. (2007). Regionalization of Municipal Solid Waste Management in Japan: Balancing the Proximity Principle with Economic Efficiency. *Environmental Management*, 40(1), 12–9.

Osumi, S. (1972). *Gomi Sensō*. Tokyo: Gakuyō Shobō.

Pastor JR., M., Sadd, J., & Hipp, J. (2001). Which Came First? Toxic Facilities, Minority Move-in, and Environmental Justice. *Journal of Urban Affairs*, 23(1), 1–21.

Pellow, D. N. (2000). Environmental Inequality Formation: Toward a Theory of Environmental Injustice. *American Behavioral Scientist*, 43(4), 581–601.

Pellow, D. N. (2004). *Garbage Wars: The Struggle for Environmental Justice in Chicago*. Cambridge: The MIT Press.

Portney, K. E. (1991). *Siting Hazardous Waste Treatment Facilities: The Nimby Syndrome*. New York: Auburn House.

Rabe, B. G. (1994). *Beyond Nimby: Hazardous Waste Siting in Canada and the United States*. Washington, DC: Brookings Institution Press.

Rootes, C. (2007). Acting Locally: The Character, Contexts and Significance of Local Environmental Mobilisations. *Environmental Politics*, 16(5), 722–741.

Rootes, C. (2009a). Environmental Movements, Waste and Waste Infrastructure: An Introduction. *Environmental Politics*, 18(6), 817–834.

Rootes, C. (2009b). More Acted upon than Acting? Campaigns against Waste Incinerators in England. *Environmental Politics*, 18(6), 869–895.

Rootes, C., & Leonard, L. (2009). Environmental Movements and Campaigns against Waste Infrastructure in the United States. *Environmental Politics*, 18(6), 835–850.

Saha, R., & Mohai, P. (2005). Historical Context and Hazardous Waste Facility Siting: Understanding Temporal Patterns in Michigan. *Social Problems*, 52(4), 618–648.

Scally, P. C., & Tighe, J. R. (2015). Democracy in Action?: NIMBY as Impediment to Equitable Affordable Housing Siting. *Housing Studies*, 30(5), 749–769.

Shibata, T. (2001a). Seijiteki Funsō Katei ni okeru Masu Media no Kinō 1. *Hokudai Hōgaku Ronshū*, 51(6), 1929–1959.

Shibata, T. (2001b). Seijiteki Funsō Katei ni okeru Masu Media no Kinō 2. *Hokudai Hōgaku Ronshū*, 52(2), 573–601.

Shimizu, S. (1999). *Nimbī Shindorōmu Kō: Meiwaku Shisetsu no Seiji to Keizai*. Tokyo: Tōkyō Shimbun Shuppan Kyoku.

Sun, Y. (2015). Facilitating Generation of Local Knowledge Using a Collaborative Initiator: A NIMBY Case in Guangzhou, China. *Habitat International*, 46, 130–137.

Taguchi, M. (2003). *Gomi Funsō no Tenkai to Funsō no Jittai*. Tokyo: Hon-no-Izumi Sha.

Takahashi, H. (2001). Gomi Kōikika o Kenshō suru. In Yamanashi Gakuin Daigaku Gyōsei Kenkyū Sentā (ed.), *Kōiki Gyōsei no Shosō* (pp. 53–92). Tokyo: Chūō Hōki Shuppan.

Takahashi, L. M. (1999). *Homelessness, AIDS, and Stigmatization: The Nimby Syndrome in the United States at the End of the Twentieth Century*. New York: Oxford University Press.

Tsugawa, T. (1993). *Dokyumento Gomi Kōjō*. Tokyo: Gijutsu to Ningen.

United Church of Christ Commission for Racial Justice. (1987). *Toxic Waste and Race in the United States: A National Report on the Racial and Socioeconomic Characteristics of Communities with Hazardous Waste Sites*. New York: United Church of Christ.

Walsh, E. J., Warland, R., & Smith, D. C. (1997). *Don't Burn it Here: Grassroots Challenges to Trash Incinerators*. University Park, PA: Pennsylvania State Press.

Wolsink, M. (2000). Wind Power and the NIMBY-myth: Institutional Capacity and the Limited Significance of Public Support. *Renewable Energy*, 21, 49–64.

Yorimoto, K. (1974). *Gomi Sensō*. Tokyo: Nihon Keizai Shimbun Sha.

Yorimoto, K. (1977). Sanka to Chiiki Seiji o meguru Gomi no Seijigaku. *Chiiki Kaihatsu*, 4, 1–57.

Young, M. G. (2012). Necessary but Insufficient: NIMBY and the Development of a Therapeutic Community for Homeless Persons with Co-morbid Disorders. *Local Environment*, 17(3), 281–293.

2 Explaining the rise and fall of the dominance of an idea

This chapter describes a theoretical framework that will be used to explain the rise and fall of an idea of distributive justice in siting policies. By overviewing studies on siting conflicts and environmental justice, the first part introduces a typology of ideas of distributive justice in siting of LULUs (Locally Unwanted Land Uses). There are multiple ideas of distributive justice and which one of them becomes dominant is crucial in siting politics. Secondly, ideational approaches in political studies are briefly reviewed and a new comprehensive framework is shown to analyse the changing influence of an idea. The third part introduces four different types of variables (i.e. ideational legitimacy, interests, power of claimants, and exogenous environments) to explain the dominance of an idea and emphasises the importance of considering not only each one of these four variables but also the interaction between them. Lastly, the methodology for this study is explained.

Conceptualising distributive justice in LULUs siting

While distributive justice is often claimed by those who bear disproportionate burdens of LULUs and incorporated in siting policies, what distributive justice means is ambiguous and even controversial. The meaning of distributive justice is open to multiple interpretations and which idea of distributive justice is dominant and adopted to siting projects affects who is to shoulder how much of the burden. Therefore, conflicts often arise over how to interpret distributive justice while distributive justice is the goal for all sides in a distributive conflict (Stone, 2001).

Researches on siting conflicts and environmental justice have shown diverse ways to conceptualise distributive justice.[1] To analyse the empirical results, this section briefly reviews the literature on distributive justice in siting conflicts by distinguishing it into four basic ideas: fair distribution of LULUs, burden/benefit balance, source reduction, and procedural fairness (Nakazawa, 2017).

Fair distribution of LULUs concerns the pattern of the distribution of LULUs and requires LULUs to be fairly distributed among communities. Approaches such as a comprehensive programme which allocates the burdens of multiple types of facilities equitably among communities are based on this idea (Gerrard, 1996; Morell, 1984). New York City's fair share approach, by which each community takes its fair share of the burden of LULUs and of beneficial public services as well, is an actual application of this approach (Rose, 1993; Valletta, 1993; Weisberg, 1993). The self-sufficiency principle in waste management is also a variation of this idea which generally requires each community to take responsibility for disposing of their own waste without imposing undue burdens on other places or people (Watson and Bulkeley, 2005).

It is noteworthy that fair distribution of LULUs differs in how, what, and among whom LULUs are to be distributed. There are various ways to distribute LULUs, ranging from simple equality, distribution proportionate to certain criteria such as needs or population, to progressive distribution in which wealthier communities host more LULUs than poor ones (Been, 1992, 1994; Deutsch, 1975; Stone, 2001). What facilities should be counted in these equations is also problematic. As there are various types of LULUs, which LULUs are counted makes a difference in assessing fairness in distribution. In addition, whether or not an outcome is seen as fair depends on how a community is defined, given the various concepts of communities in the siting of a project (Lesbirel, 2011b). As the controversy over appropriate geographical scales to assess environmental inequality indicates, fairness based on one definition of community does not necessarily guarantee fairness based on another definition (Mohai, Pellow & Roberts, 2009).

Burden/benefit balance requires that the inherent discrepancy of burden and benefit caused by a locally unwanted facility should be balanced. This idea includes two approaches: compensation and burden minimisation. Compensation aims to set off negative impacts of LULUs by providing positive goods (Kunreuther, 1986; O'Hare, 1977), which include not only monetary compensation, but also non-monetary compensation such as providing locally beneficial facilities. Burden minimisation is an approach which aims to redress the imbalance by mitigating the burdens and risks caused by LULUs through, for example, pollution control technologies, monitoring the operation of the facility and so forth (Mazmanian & Morell, 1992; Portney, 1985, 1991).

Source reduction is an idea to achieve distributive justice by minimising waste production at the source and reducing the number or scale of LULUs. While the fair distribution of LULUs and burden/benefit balance takes the necessity of LULUs for granted and focuses on how to distribute necessary facilities, the idea of source reduction is oriented to address root causes which necessitate LULUs. This view of distributive justice is increasingly common, as sustainable development becomes a dominant discourse. Environmental justice has broadened its perspective from fair distribution of environmental hazards to control over production of environmental hazards at source, as symbolised in *Not In Anybody's Back Yard* (Boudet, 2011; Burningham, Barnett, & Thrush, 2006; Futrell, 2003; Glasgow, 2005; Heiman, 1990). It is being increasingly advocated among local and environmental activisms that more effort should be concentrated on source reduction rather than constructing new facilities.

Procedural fairness pertains to the way in which collective decisions on the distribution are made. Distributive justice is not only about the end result, but also about the process by which the result occurs (Stone, 2001). From this perspective, a distribution is justifiable if it results from a fair process. Scholars and practitioners of siting increasingly turn to more democratic and voluntary processes, recognising that Decide-Announce-Defend approaches do not work well. Environmental justice studies also point out that fair process is indispensable to realising distributive justice, empowering the least-empowered and opening up the way to change the systems which reproduce distributive injustice (Heiman, 1996; Hunold & Young, 1998; Lake, 1993, 1996).

The idea of fairness in siting process is also diverse, ranging from impartiality, equal opportunity, equal chance, voluntariness, to participation. Impartiality requires that a site selection be made on unbiased, non-partisan criteria. A process is fair when sites for LULUs are chosen according to technical and rational criteria, and all communities are treated equally without any discriminatory intention. Equal opportunity argues for fair competition among communities with equal starting resource, such as competitive bidding against LULUs among communities with an equal number of bargaining chips such as vetoes (Been, 1992). Although monetary auction schemes also emphasise fair competition, they are weak in terms of equal opportunity given inequalities of wealth that existed prior to the auction (Been, 1992). They are usually complemented with equal chance or voluntariness (Inhaber, 1998; Kunreuther, 1996). Equal chance requires that each community have an equal chance of being selected for the site. A typical example is a lottery (Kunreuther, 1996; Oberholzer-Gee, Bohnet, & Frey, 1997). Voluntariness emphasises a community's willingness to accept LULUs (Munton, 1996). A process is fair if communities are given veto power and make their own decisions on whether or not to host LULUs, usually on certain conditions such as compensation schemes, risk mitigation measures and the like. Because voluntariness leaves open the question on how a community decides to volunteer, this approach is often combined with participation (Munton, 1996; Rabe, 1994). Participation demands that all parties are equally represented in a process. It values informed citizens' communication to assess possible alternatives, arguing that ordinary citizens are capable of deliberating and making a sensible decision (Kraft & Clay, 1991; Hunter & Leyden, 1995). Post positivism viewpoints also advocate a participatory deliberative process to reconcile different, conflicting views and values which those involved in siting conflicts hold (Fischer, 1993, 2000; McAvoy, 1999; Renn, Webler, & Kastenholz, 1996; Renn, 2006).

Thus, there are multiple ideas of distributive justice, while it is a significant value in siting conflicts. Although they are not necessarily mutually exclusive and usually several ideas are combined in an approach, which one of them dominates policies makes a significant difference in the distribution of the burden. In-Ward Waste Disposal (IWWD), which requires incinerators to be fairly distributed among the wards, is a variation of the first type of distributive justice. As delineated in the introduction, this idea of distributive justice has been recognised as a significant principle of waste management in Japan, but its influence on siting policies of Tokyo fluctuated over time. Why does a particular idea of distributive justice become more influential at a particular period of time? Conversely, why does a once-influential idea decline and eventually get abandoned? To explain the rise and fall of the dominance of an idea, this book turns to ideational approaches in political studies.

Ideational approaches in political studies

Significance of ideas to explain political phenomenon

An increasing number of scholars have paid close attention to the role and influence of ideas in shaping political outcomes (Béland & Cox, 2010; Bell, 2012; Berman, 2001; Blyth, 2013; Campbell, 1998, 2002; Padamsee, 2009; Panizza & Miorelli, 2013; Schmidt, 2008; N. Smith et al., 2014; Surel, 2000). They acknowledge that ideas, such as conceptual models, norms, values, world views, frames, discourses, causal beliefs, cultures, ideologies, and the like, affect political behaviours and outcomes. In general, ideas shape "how we understand political problems, give definition to our goals and strategies, and are the currency we use to communicate about politics" (Béland & Cox, 2010: xvi). Scholars in ideas and politics have shown that ideas are not epiphenomenal or mere political hooks on material interests, but factors which exert significant causal influence on politics and policy making. By focusing on ideas, it is possible to explain political behaviours and outcomes which cannot be explained with rationalist interests, material conditions, institutional settings, or historical paths (Schmidt, 2008, 2010a).

Types and roles of ideas discussed in these studies vary from specific, concrete, programmatic ideas to broader, more general ideas (Tannenwald, 2005). Policy-relevant ideas and their roles in politics can be differentiated according to levels of generality (Campbell, 1998, 2002; Mehta, 2010; Schmidt, 2008). At the most concrete and specific level, ideas can take the form of specific policies or solutions proposed by policy makers. Ideas at this level provide the means for solving problems and accomplishing objectives. Programmatic ideas such as policy paradigms (Hall, 1993) and policy definitions (Mehta, 2010) are at a more basic level than policy ideas. They define the problems to be solved, the goal to be achieved, and the norms to be applied by such policies. Philosophy, or worldviews, refers to broader ideas which cut across substantive areas and undergird the other two levels of ideas. Policy and programmatic ideas are seen as being at the foreground, and are discussed and debated on a regular basis. In comparison, philosophical ideas – as a set of taken-for-granted assumptions which are widely shared – are not open to criticism, being at the background. Ideas also differ in whether they are normative or cognitive (Campbell, 1998, 2002; Schmidt, 2008). Normative ideas consist of values and attitudes which inform what is legitimate, appropriate and what one ought to do. Cognitive ideas are descriptions and theoretical analyses which specify causal relationships and provide guidelines and roadmaps for political actions.

The role and influence of ideas have been discussed and applied in various fields of social sciences. Some scholars in institutionalism incorporate ideational factors to explain policy changes and political outcomes which cannot be explained with cultural, historical and rational choice institutionalisms

(Béland, 2005, 2009b; Béland & Waddan, 2015; Blyth, 1997, 2001; Campbell, 1998, 2002; Hall, 1993, 1996; Moon, 2013; Peters, Pierre, & King, 2005; Schmidt, 2002a, 2002b, 2003, 2008, 2010a, 2010b). Comparative policy analysis also takes ideational factors seriously to explain variations in specific policies among developed democracies (Bleich, 2002; Cox, 2004; King, 1973; Schmidt, 2002a, 2002b; Skogstad, 1998; Walsh, 2000; White, 2002). Constructivists in international relations emphasise the role of shared ideas in foreign policy (Haas, 1992; Parsons, 2000; Wendt, 1999). Studies focusing on racial and gender politics also have paid close attention to ideational processes in the formation of identities and inequalities (Béland, 2009a; Bleich, 2002; King & Smith, 2014; Lieberman, 2002, 2010; Marshall, 2000; Padamsee, 2009; Van Dijk, 1992; White, 2002; Whitehead and Tsikata, 2003). In the studies on environmental politics and disputes, the impact of ideas, discourses and arguments have been discussed as different perceptions and perspectives on environmental issues compete with each other (Dryzek, 2005; Fischer, 2000; Hajer, 1997; Hajer & Versteeg, 2005; Litfin, 1994; Teravainen, 2010). Studies on social movements and mobilisation stress the significance of ideologies and discursive frames in mobilisation processes, challenging materialist approaches (Benford, 1993; Benford & Snow, 2000; Snow et al., 1986).

Studies on the siting of a project have also paid close attention to the ideational aspects of conflicts. From this perspective, siting conflicts are viewed as conflicts between different beliefs, ideas or discourses around key aspects of siting (Lesbirel, 2011a). In siting conflicts, there is difference among relevant actors in the very way the problem is understood and defined, as siting issues are wicked problems with no clear-cut criteria by which a resolution can be judged (Fischer, 1993, 2000). The tension between citizens and experts in siting conflicts is rooted in their differing point of view in relation to key issues in siting (McAvoy, 1999), such as the division in risk perception (Cvetkovich & Earle, 1992). More democratic and deliberative processes are advocated to reconcile different perspectives, as there is no objective, scientific standpoint to evaluate them (Fischer, 2000; McAvoy, 1999). Ideational viewpoints are also applied to studies which attempt to reveal how specific identities or stigmas are mobilised in siting conflicts, particularly ones over human service facilities. For instance, Hubbard (2005) and Wilton (2002) argue that the notion of "Whiteness" is mobilised in community opposition to human service facilities to defend socio-spatial privileges of the white population. The role of stigma is also explored to explain the community rejection of human services related to homelessness and HIV/AIDS (Takahashi, 1997a, 1997b; Takahashi & Dear, 1997). Framing analysis, a social constructionist approach in social movement theory, is applied to grassroots movements against LULUs as well. Framing refers to the actors' signifying work or construction of meaning (Benford & Snow, 2000; Snow & Benford, 1988; Snow et al., 1986). Framing analysis studies sets of beliefs and meanings which inspire and legitimise the activities and campaigns of a social movement. Some studies on siting take this approach and illustrate how grassroots movements against LULUs expand their narrow, reactive goals through interaction with the opponents and nationwide movement networks (Futrell, 2003; Kang & Jang, 2013; Mcclymont & O'hare, 2008; Shemtov, 2003).

Explaining the dominance and its change over time

This book attempts to examine why a particular idea of distributive justice becomes dominant in siting policies and why its dominance changes over time. Most of the studies in ideas and politics have aimed to prove the ideas' causal influence on political phenomena. However, the purpose of this book is not to show whether ideas matter or not in politics, but to examine why a particular idea rises in prominence in policies. Ideas matter in politics and policy making. Then, why does a particular idea, not others, thrive at a particular period of time? This question indicates treating an idea as what needs to be explained rather than the explanation (Acharya, 2004; Béland & Cox, 2010). Ideational researchers have been more interested in treating an idea as the explanation to demonstrate that ideas are major causal factors in explaining politics and policy making, thus refuting materialist approaches (Berman, 2001). While a number of theories and case studies show that an idea is important and influential in explaining specific political and policy outcomes, why particular ideas are successful among multiple competing ideas has been much less studied (Mehta, 2010).

Furthermore, this study is interested not only in the dominance of an idea in one period of time, but also in the changes of that dominance from one period to another. The degree to which an idea is reflected in policies fluctuates over time. The same idea may be influential on policies in one period, but less dominant in another. To grasp an overall picture of the politics of ideas, it is necessary to explain not just when an idea's time comes but also when that idea's time is up (Mehta, 2010). While attention has been paid to the former, the latter has been much less studied (Mehta, 2010; Schmidt & Radaelli, 2004). Although it has been pointed out that the failure of previous ideas opens up space for a new idea to come in (Berman, 2001, 2010; Campbell, 2002; Goldstein & Keohane, 1993; Legro, 2000), what makes a once-dominant idea become less influential and eventually fail has been understudied.

To explain the dominance of an idea and its changes over time, ideational explanation alone is not sufficient. It is necessary to have an approach that integrates ideational and other types of variables which shape the production of particular ideas and the conditions under which they can affect behaviours and outcomes (Béland, 2009b; Béland & Cox, 2010; Berman, 2001, 2010; Florini, 1996; Padamsee, 2009; Schmidt & Radaelli, 2004; Walsh, 2000). Non-ideational factors play significant roles in determining which idea is victorious. This is not to disregard the influence of ideational factors. Ideas may become prominent and powerful more or less in their own right and/or its resonance with pre-existing dominant ideas at a more general level may make them influential. However, ideational factors alone cannot explain why a particular idea, among multiple competing ideas, achieves prominence in the political realm at particular moments while others do not (Berman, 2001; Lieberman, 2002, 2010; Schmidt & Radaelli, 2004).

Politics is both powering and puzzling (Hall, 1993; Heclo, 1974; Lieberman, 2010). Interests and power of actors must play significant roles in whether an idea becomes successful or not in policies. Ideas need the support of powerful actors that have an interest in promoting them (Béland, 2005). External events and environments should also be taken into account, for an idea becomes prominent under specific environments and any change in them is likely to affect the influence of that idea. To answer the question of the rise and fall of an idea, it is necessary to take a comprehensive, integrated framework which recognises not only ideational causes but also the power struggles between rationally calculating actors, as well as the influence of external events and environments (Nakazawa, 2016).

The dominance of an idea and its four explanatory variables

This study, therefore, examines different types of variables and the interaction between them to explain the dominance of an idea. This section explains variables to be considered in the empirical analysis. Firstly, it clarifies what is to be explained (i.e. the dependent variable) and how it is measured. Then, four variables (i.e. ideational legitimacy, interests, power of claimants, and exogenous environments) are introduced to explain the dominance of an idea. It is argued that it is important to look into not only the effects of each one of these four variables, but also the way in which they interact with one another, in order to reveal mechanisms behind the diachronic changes of the dominance of an idea.

- The dependent variable
    — The dominance of an idea: the degree to which an idea is reflected in policies
- The explanatory variables
    — Ideational legitimacy: the degree to which an idea is perceived as normatively and cognitively legitimate
    — Interests: the degree to which an idea fits with interests among actors, helping them achieve their goals
    — Power of claimants: the ability of the carriers of an idea to influence policy processes and outcomes
    — Exogenous environments: the degree to which an idea fits with environments outside of the concerned political arena

Defining the dominance of an idea

The dependent variable in this research is the dominance of an idea, herein defined as the degree to which that idea is reflected in governmental policies. It is possible to define the dominance of an idea in other ways, given the variety of ways in which ideas influence political thoughts, behaviours and outcomes as reviewed above. It may be defined, for instance, as the degree to which an idea is shared among social groups such as policy elites, politicians, academics, social movements, general citizens, and the like. It is also possible to operationalise it as the degree to which an idea is prevalent in mass media, books, or the internet. However, this study focuses on the impact of an idea on governmental policies. Being prevalent among certain social groups and/or media does not necessarily guarantee that an idea dominates distributive issues in politics. For an idea of distributive justice to influence outcomes, that idea must be substantively adopted in governmental policies. Focusing on governmental policies offers an appropriate platform to evaluate the dominance of an idea of distributive justice.

In this study, the influence of an idea in policies is measured by two criteria:

- whether that idea is recognised as a significant principle in policies,
- and the degree to which that idea substantively affects contents of policies.

An idea is dominant when policies recognise that idea as a central principle. In other words, the dominance is measured according to whether an idea is mentioned or not in policy documents. However, this simple measurement alone is not sufficient to assess the dominance of an idea, because the fact that an idea is mentioned in policy documents does not guarantee that the idea is substantively reflected in policies. It is possible that an idea may be compromised and does not exert substantial impact on policies even if it is worded in documents. It is necessary to look into whether an idea actually influences the substantive contents of the policies. For this purpose, the study sets as the second criterion the degree to which the requirements of an idea are substantively embodied in the policies. For instance, an idea that requires siting an incinerator in every city is dominant in a policy when that policy actually shows concrete plans to construct an incinerator in every city. Conversely, its influence is relatively weak, regardless of whether that idea is advocated in that policy, if that policy does not show any clear roadmap to satisfy what that idea requires and/or that requirement is compromised allowing one incinerator to be shared with several cities.

This does not mean excluding the enforcement and implementation of the policy from the analysis. While the book assesses the dominance of an idea by the degree of its prevalence in policies rather than its impacts on the resultant distribution, it takes into consideration the feedback from the implementation and its effects on the dominance of an idea in policies. While scholars in politics and ideas have paid much attention to the adoption of new ideas, the implementation of such adopted ideas has been less focused upon (Schmidt & Radaelli, 2004; Walsh, 2000). Nonetheless, the implementation plays a significant part in the story of the success or failure of ideas. It is all the more true for LULUs' siting because the difficulty of facility siting lies in most part in its implementation. Conflicts happen not only in the acceptance of an idea into policies, but also in the process of carrying out the siting of a project to realise that idea. Furthermore, this feedback from the implementation is particularly important as this study is interested not only in the dominance of an idea at one period of time, but also its change over time. Even if policy makers once accept an idea, it may be compromised later when faced with the difficulty to put it into practice. Although this study focuses on the dominance on an idea "in policies", it considers the implementation of the policies as a significant factor.

Ideational legitimacy

Ideational legitimacy concerns the characteristics of an idea which are perceived as legitimate among actors. As framing theory contends, ideas mobilise potential adherents and constituents, garner bystanders' support, and demobilise antagonists (Snow & Benford, 1988). How an idea is associated with elements, such as causality of the problems, culpable actors, victimisation, solutions, cultural symbols, and values, affects the likelihood of the success of that idea (Benford & Snow, 2000; Bleich, 2002; Campbell, 1998; Mehta, 2010; Schmidt, 2008; K. E. Smith, 2013; Snow & Benford, 2000; Snow et al., 1986; Stone, 1989, 2001). In other words, the contents of an idea affect its chances of dominating policies.

The dominance of an idea depends on whether it satisfies the underlying values of the polity which policy makers and citizens share (normative) and whether it provides robust solutions to existing problems (cognitive) (Schmidt, 2008). An idea of distributive justice contains both normative and cognitive arguments as well. An idea of distributive justice in siting is normative as it informs how the burden of locally unwanted facilities ought to be distributed among communities and social groups. The chance of an idea dominating policies is high when its normative implications are morally appealing. An idea of distributive justice has cognitive arguments as well, specifying causal relationships and providing guidelines and roadmaps for political action. An idea is more likely to be adopted by actors if it is perceived as a viable solution to existing problems, providing a clear roadmap out of troublesome policy situations (Berman, 2001; Campbell, 1998, 2002; Hall, 1989; Schmidt, 2008; Schmidt & Radaelli, 2004). An idea becomes dominant when it is normatively appealing and cognitively convincing, and when normative and cognitive arguments are compatible rather than contradictory (Schmidt & Radaelli, 2004).

It is noteworthy that ideational legitimacy is not just about internal structures and logic of ideas, but also about its congruence with pre-existing normative and cognitive frameworks. The internal characteristics alone cannot explain the variation in the degree of ideational legitimacy among places and times. It is necessary, therefore, to look into the relationship with pre-existing and prevailing normative and cognitive frameworks. A new idea is more likely to prevail when it is congruent with such frameworks which affect how actors think and behave (Acharya, 2004; Campbell, 1998, 2002; Mehta, 2010; Schmidt, 2010a; Schmidt & Radaelli, 2004; A. Smith & Kern, 2009; K. E. Smith, 2013; White, 2002). An idea of distributive justice is perceived as legitimate when its normative arguments resonate with actors' values and its cognitive arguments fit with the existing policy paradigm (Hall, 1993), or policy core (Sabatier, 1988).

For instance, an idea of distributive justice is more likely to be accepted in policy making when that idea's cognitive arguments fit with a prevailing policy paradigm. A policy paradigm is an interpretive framework of ideas and standards that specifies not only the goals of policy and the kind of instruments that can be used to attain them, but also the very nature of the problems they are meant to be addressing (Hall, 1993). Policy paradigms constrain actors' thoughts and practices by specifying goals and limiting the range of instruments that can be used. An idea may be perceived as a viable solution to existing problems under a certain policy paradigm, and the shift of that policy paradigm is likely to affect the dominance of a particular idea.

Interests

Interests (material benefits, political powers, or others) are also a significant factor that can be used to explain the dominance of an idea. An idea is more likely to be prominent in policies when that idea is congruent with the interests of relevant actors, helping them further and achieve their self-interested agendas. As many researchers argue, interests are important conveyors of ideas (Carstensen, 2010, 2011; Dudley, 1999; Goldstein & Keohane, 1993; Hall, 1989; Hansen & King, 2001; Howorth, 2004; Stone, 1993; Surel, 2000; J. I. Walsh, 2000). Even if an idea is perceived as ideationally legitimate, actors may not support that idea if it is against their interests. On the contrary, a bad idea may prevail if it is congruent with the key interests of important actors. The more an idea, or the policy implications of that idea, fits with the strategic interests among actors, the more likely that idea becomes influential on policies.

It is at least partly true that even interests are socially constructed through ideational processes, as constructivism approaches maintain (Béland, 2007; Cox, 2001; Dimitrakopoulos, 2005; Hay, 2010; Hofmann, 1995; Padamsee, 2009; Schmidt, 2010a; White, 2002). Constructivists criticise the pluralist and rationalist views on ideas as being epiphenomenal and mere hooks for material interests. They insist that interests are shaped through ideational processes. As Schmidt (2010a) summarises, even where focus is on strategic and instrumental behaviour, the emphasis is on agents' ideas about their subjective interests, about which utility to maximise (interests), how to maximise it (strategies) and to what end (goals). Separation of ideas and interests in the rationalist framework leads to treating ideas as mere justification or legitimisation for interests (Laffey & Weldes, 1997).

Nonetheless, it is convenient and reasonable to analytically separate interests and ideas, for ideas and self-interests often contradict each other in siting disputes. It is common for all parties to agree on an idea in general, but some oppose a concrete plan when it imposes the burden of LULUs on their community. Siting conflicts are not just about beliefs, norms and values, but also about gain and loss. A new idea of distributive justice, if it is adopted in policies, changes the pattern of the distribution of costs and benefits among communities and groups. It is likely that those who are expected to accept burdens under a new idea oppose that idea even when they consider it normatively and/or cognitively legitimate. Politics of distributive justice in siting are more subject to the struggle over interests. It is reasonable and productive to analytically separate ideational causes and interests, and empirically examine how they interact in the political processes.

Power of claimants

Power concerns the ability of the claimants of an idea to influence policy-making processes and outcomes. New ideas do not achieve political prominence on their own, but must be championed by carriers (or advocates) such as entrepreneurs, policy makers, politicians, social groups and movements, who are able to make others reconsider the ways they think and act (Béland, 2005, 2009b; Berman, 2001; Florini, 1996; Hansen & King, 2001; Howorth, 2004). There will be proponents and opponents of an idea, whether from ideational- or interest-oriented reasons, and both sides fight over that idea in the policy-making processes. Ideational legitimacy and interests give actors motivation, but not their ability to make it prevalent. Politics involves power; an idea, however good or compelling in its arguments, can nevertheless fail if certain actors with veto power remain unconvinced (Gourevitch, 1989; Schmidt & Radaelli, 2004). For an idea of distributive justice to become influential in policies, the claimants of that idea need to be capable of making their claims heard by the policy makers.

Power also includes the capacity of a government to implement the policy requirements of an idea. In this regard, power refers to not only the ability to override the opposition. As Hall (1989) contends, administrative feasibility affects the success or failure of a policy idea. Even if an idea of distributive justice is adopted by the government, the application to policies may be compromised when that idea is not feasible given insufficient governmental capacity. For instance, an idea may lose its dominance in policies when the financial cost of realising that idea is beyond governmental capacity to fund it. Not only financial capacity, but also urban structure, such as land availability, traffic conditions, population distribution, or ethnic and class divisions, may constitute major constraints on possible siting options and undercut the governmental capacity to put that idea into practice.

Power, herein defined as the ability of proponents of an idea to influence policy processes and outcomes, is held by various sources. Organisational resources constitute a significant repository of power. Economic resources, human resources, material forces and such help actors who possess them to make their opinions heard in policy-making processes. As many researchers argue, institutional settings (the formal and informal rules and procedures in policy making) also affect which ideas are carried to the policy-making arena, and thereafter adopted and implemented as a policy (Béland, 2005, 2009b; Campbell, 1998, 2002; Dimitrakopoulos, 2005; Lieberman, 2002, 2010; Mehta, 2010; Walsh, 2000). For instance, in siting conflicts, local vetoes to the siting of LULUs may limit the power of those who advocate an idea of distributive justice. On the other hand, authoritative measures to overcome opposition, such as eminent domain, may empower government officials to push forward with an idea of distributive justice more effectively. Institutions, as a condition for political decision making and implementation which distribute power and resources between actors, are a major determinant of the extent of the actors' power as well as their ability to wield that power (Hall, 1989).

Exogenous environments

Exogenous environments refer to factors that are external to the relevant political system, such as fundamental socio-cultural values and social structures, demographic conditions, economic situations, technological development, nationwide governing coalition, and so forth. Not only factors within a political system, but also those outside of that system can affect the policy-making process and thereafter the likelihood of an idea becoming dominant in the policies. For instance, the policy-making process in a local political system is likely to be influenced by changes and incidents at the national and/or international level, which are normally beyond the control of those within that political system. To understand the dominance of an idea and its variation over a long period of time, factors outside of the local political system should be taken into account.

As many researchers maintain, the contingency with the external environments affects the chances of an idea being successful (Béland & Orenstein, 2013; Berman, 2001; Campbell, 2002; Florini, 1996; Sabatier, 1988; Schmidt & Radaelli, 2004; Surel, 2000). For instance, economic changes may impact the siting of facilities when the demand for them is significantly dependent on economic circumstances. Technological innovation may make a certain technology obsolete and discredit an idea of distributive justice if it is closely related to that outmoded technology. Governmental changes in central politics may change the relationship among actors in the concerned local political system. By the same token, global financial crises, wars, or international treaties also might impact the dominance of an idea in one way or another. Not only material factors such as economic conditions or technology, but also ideational environments matter as well. Pre-existing ideas which are prevalent beyond the concerned political system, such as worldviews, cultural values, or long-held policy paradigms, provide environments which affect how actors in the political system think and behave in regard to an idea of distributive justice. Thus, the inputs from environments external to the concerned political system are a significant factor that can be used to explain the dominance of an idea, as they impose or provide, as the case may be, constraints and resources to the policy-making processes.

The interaction between the variables

This book argues that the dominance of an idea in policies at any one period of time is determined by the interaction between these four explanatory variables. The four variables are not independent of each other. Rather, they interact in the policy-making processes. Examining the effects of each variable is not sufficient to explain the rise and fall of the dominance of an idea; it is also necessary to understand the way in which they interact with one another and how such interaction results in the dominance, or the decline, of an idea. As Rihoux and Ragin argue, "the idea that each single cause has its own separate, independent impact on the outcome should be replaced by the assumption that 'conjunctural causation' is at work" (Rihoux & Ragin, 2009: 9); multiple causes can be combined and constitute a causal combination.

This book, therefore, focuses on the interaction between the four variables. There might be synergies and/or conflicts between the variables. For instance, ideational legitimacy and self-interests may work harmoniously and give a boost to an idea. It is also probable that one variable conflicts with another. Ideational legitimacy may override entrenched interests of actors when an idea provides a viable solution to existing problems and satisfies the actors' underlying values. Conversely, individual interests of powerful actors may prevent an idea from influencing policies even if that idea is perceived as normatively and cognitively legitimate by all of the parties. Even if the exogenous environments are not hospitable to an idea, that idea may still prevail, if it is backed by some actors who are able to push it into policies.

More importantly, there might be interrelations between the variables. It is possible that one variable influences another through some intermediate mechanisms. For instance, a certain exogenous environment might affect the ideational legitimacy of an idea, positively or negatively, if the idea is believed to provide a viable solution to the problem produced by that environment. By the same token, the actors' power may be related to economic or political conditions outside of the concerned political system. It is also likely that a dissonance of interests within a group will weaken the power of that organisation. To understand the causes and mechanisms behind the dominance of an idea, examining each of the four variables will not be sufficient; it is necessary to unravel how the four variables interact with one another and how that interaction affects the dominance of an idea in policies.

Changes in the dominance of an idea

This book explains not only the dominance of an idea at one period of time, but also its changes over time. As noted earlier in this chapter, while much more attention has been paid to why an idea's time comes, why an idea's time is up has been understudied. To grasp a more comprehensive picture of the politics of ideas, it is essential to illuminate mechanisms by which the dominance of an idea varies from one period to another.

This book argues that the dominance of an idea changes over time as the four variables and the way they interact change from one period to the next. As many scholars argue, changes may be brought about by external shocks (Berman, 2001, 2010; Campbell, 2002; Florini, 1996; Moschella, 2015; Sabatier, 1988; Schmidt & Radaelli, 2004; Surel, 2000). Changes in environments external to the political system, such as economic crises, technological innovation, ideological shifts, governmental changes, or international conflicts, could increase or reduce the influence of an idea on policies. Consequently, analysing the long-term variation in the dominance of an idea requires paying close attention to the changes in exogenous environments.

However, the external shocks are not the only cause of changes. Even when the environments do not change, the dominance of an idea may still change as the other explanatory variables change. Policy changes can happen endogenously without external shocks (Lieberman, 2002). The power of proponents might change from one period to another, as the power relationship among actors changes through some mechanisms within the political system. Likewise, an idea's congruence with interests may change as institutional settings in the political system change. Policy elites' perception of the ideational legitimacy of an idea may change as a result of their policy-oriented learning[2] (Hall, 1993; Heclo, 1974; Sabatier, 1988). To grasp the causes behind the changes in the prominence of an idea, it is necessary to pay attention to the changes in both the exogenous and the endogenous factors.

Furthermore, it is necessary to take into consideration not only the changes in each one of the variables but also in the way they interact. As noted above, the variables are not independent, and they interact with one another. The relation between the variables may change if one of them changes. For instance, growing normative legitimacy of an idea may persuade actors who formerly opposed it, due to incongruence with their interests, to now accept it. Moreover, a change in a variable may induce change of another. For instance, a change in the power relation might lead to a change in political interests among actors, or vice versa. To explain the changes in the dominance of an idea, just examining the changes in each one of the variables is not sufficient; it is necessary to reveal how a change in one variable affects the way the variables interact and how it results in the change in the dominance of an idea.

Methodology

This research conducts a diachronic comparative case study on IWWD in the 23 wards of Tokyo. An in-depth qualitative case study informs this project. This book attempts to unravel the political processes in which an idea of distributive justice waxes and wanes by examining the four explanatory variables and their interaction. This requires an in-depth investigation into the relevant actors' motives and behaviours in the policy-making and implementation processes which are also embedded in unique political, material, and ideational contexts inside and outside of the concerned political system. A case study offers "a means of investigating complex social units consisting of multiple variables of potential importance in understanding the phenomenon" (Merriam, 1988: 41).

Diachronic comparison is another feature of this project. Comparing different periods of time can offer a better understanding of what determines the dominance of a particular idea of distributive justice and why it changes over time. IWWD in the 23 wards of Tokyo provides an interesting case, as the dominance of this idea of distributive justice varied through the four periods as explained in the previous chapter. The comparison of these four periods can strengthen the argument on the causal mechanisms of the dominance of an idea.

The dominance of an idea and the impact of the four explanatory variables are assessed qualitatively with document analysis, being supplemented by semi-structured interviews. Document analysis is a systematic procedure for reviewing or evaluating documents (Bowen, 2009). It is particularly applicable to qualitative case studies by helping researchers to produce rich descriptions of socio-political phenomena, uncover meaning, develop understanding and discover insights to the issues and problems they research (Stake, 1995; Yin, 2009). This book is a case study on siting policies in Tokyo and looks into the historical changes of the dominance of IWWD in them. Documents are the primary data source for historical studies and ideational approaches. By analysing documents, researchers can track the historical change and development of policies in relation to the influence of ideas, values, and beliefs (Bowen, 2009; Yin, 2009). Main documents used in this study include: governmental policy documents, reports by governments or governmental committees, minutes of the local assemblies, surveys and opinion polls, leaflets and bulletins issued by relevant actors, judicial records, and news articles (Asahi, Yomiuri, Mainichi, Nikkei, Tosei Shimpō, and other local and trade papers).

Semi-structured interviews with those involved in the case complement document analysis. The interviews are semi-structured because questions should be tailored to specific contexts and situations in which the participants took part and how they understand distributive justice. Sixteen interviews were conducted between October 2011 and January 2012. The interviewees include; an official of the Tokyo Metropolitan Government in the 1970s, officials of the Clean Association of Tokyo 23 Wards, government officials and council members of Koto ward and other wards, members of the Tokyo Cleaning Workers Union, and leaders of local opposition movements in the 1980s and the 1990s.

Conclusion

To answer the question of the rise and fall of an idea, this project uses a comprehensive, integrated framework, which recognises not only ideational legitimacy but also the self-interests and the power of actors as well as inputs from the exogenous environments. It is argued that the dominance of an idea at one period of time is determined by the interaction between the four explanatory variables and that the dominance changes as each one of them and how they interact evolve over time. By conducting an in-depth qualitative case study on IWWD with this theoretical framework, this project aims to grasp a more comprehensive picture of the politics of distributive justice in siting conflicts.

The following four chapters analyse the case of IWWD in the 23 wards of Tokyo period by period. By comparing these four periods, this book attempts to identify what caused the rise and fall of this idea of distributive justice.

Notes

1 For instance, Been (1992, 1994) explored what fairness could mean in the context of environmental justice and showed possible different notions of fairness; LULUs are evenly distributed among communities; communities who have LULUs are compensated from those who do not; all communities receive an equal number of vetoes; those who benefit from LULUs bear them; wealthier communities host more LULUs than

poor ones; the siting process involve no intentional discrimination against people of colour; equal concern and respect for all neighbourhoods.
2 Policy-oriented learning refers to "relatively enduring alterations of thought or behavioral intentions which result from experience and which are concerned with the attainment (or revision) of policy objectives" (Sabatier, 1988: 133).

References

Acharya, A. (2004). How Ideas Spread: Whose Norms Matter? Norm Localization and Institutional Change in Asian Regionalism. *International Organization*, 58(2), 239–275.

Been, V. (1992). What's Fairness Got to Do with it? Environmental Justice and the Siting of Locally Undesirable Land Uses. *Cornell Law Review*, 78, 1001–1085.

Been, V. (1994). Conceptions of Fairness in Proposals for Facility Siting. *Maryland Journal of Contemporary Legal Issues*, 5, 13–24.

Béland, D. (2005). Ideas and Social Policy: An Institutionalist Perspective. *Social Policy & Administration*, 39(1), 1–18.

Béland, D. (2007). The Social Exclusion Discourse: Ideas and Policy Change. *Policy & Politics*, 35(1), 123–139.

Béland, D. (2009a). Gender, Ideational Analysis, and Social Policy. *Social Politics: International Studies in Gender, State & Society*, 16(4), 558–581.

Béland, D. (2009b). Ideas, Institutions, and Policy Change. *Journal of European Public Policy*, 16(5), 701–718.

Béland, D., & Cox, R. H. (2010). Introduction: Ideas and Politics. In D. Béland & R. H. Cox (eds.), *Ideas and Politics in Social Science Research*. New York: Oxford University Press.

Béland, D., & Orenstein, M. A. (2013). International Organizations as Policy Actors: An Ideational Approach. *Global Social Policy*, 13(2), 125–143.

Béland, D., & Waddan, A. (2015). Breaking Down Ideas and Institutions: The Politics of Tax Policy in the USA and the UK. *Policy Studies*, 36(2), 1-20.

Bell, S. (2012). The Power of Ideas: The Ideational Shaping of the Structural Power of Business. *International Studies Quarterly*, 56(4), 661–673.

Benford, R. D. (1993). "You Could Be the Hundredth Monkey": Collective Action Frames and Vocabularies of Motive within the Nuclear Disarmament Movement. *The Sociological Quarterly*, 34(2), 195–216.

Benford, R. D., & Snow, D. A. (2000). Framing Processes and Social Movements: An Overview and Assessment. *Annual Review of Sociology*, 26, 611–639.

Berman, S. (2001). Ideas, Norms, and Culture in Political Analysis. *Comparative Politics*, 33(2), 231–250.

Berman, S. (2010). Ideology, History and Politics. In D. Béland & R. H. Cox (eds.), *Ideas and Politics in Social Science Research*. New York: Oxford University Press.

Bleich, E. (2002). Integrating Ideas into Policy-Making Analysis: Frames and Race Policies in Britain and France. *Comparative Political Studies*, 35(9), 1054–1076.

Blyth, M. M. (1997). "Any More Bright Ideas?" The Ideational Turn of Comparative Political Economy. *Comparative Politics*, 29(2), 229–250.

Blyth, M. M. (2001). The Transformation of the Swedish Model: Economic Ideas, Distributional Conflict, and Institutional Change. *World Politics*, 54(1), 1–26.

Blyth, M. M. (2013). Paradigms and Paradox: The Politics of Economic Ideas in Two Moments of Crisis. *Governance*, 26(2), 197–215.

Boudet, H. S. (2011). From NIMBY to NIABY: Regional Mobilization against Liquefied Natural Gas in the United States. *Environmental Politics*, 20(6), 37–41.

Bowen, G. A. (2009). Document Analysis as a Qualitative Research Method. *Qualitative Research Journal*, 9(2), 27–40.

Burningham, K., Barnett, J., & Thrush, D. (2006). *The Limitations of the NIMBY Concept for Understanding Public Engagement with Renewable Energy Technologies: A Literature Review*. A Working Paper of the Research Project "Beyond Nimbyism: A Multidisciplinary Investigation of Public Engagement with Renewable Energy Technologies".

Campbell, J. L. (1998). Institutional Analysis and the Role of Ideas in Political Economy. *Theory and Society*, 27(3), 377–409.

Campbell, J. L. (2002). Ideas, Politics, and Public Policy. *Annual Review of Sociology*, 28(1), 21–38.

Carstensen, M. B. (2010). The Nature of Ideas, and Why Political Scientists Should Care: Analysing the Danish Jobcentre Reform from an Ideational Perspective. *Political Studies*, 58(5), 847–865.

Carstensen, M. B. (2011). Ideas are not as Stable as Political Scientists Want Them to be: A Theory of Incremental Ideational Change. *Political Studies*, 59(3), 596–615.

Cox, R. H. (2004). The Path-dependency of an Idea: Why Scandinavian Welfare States Remain Distinct. *Social Policy and Administration*, 38(2), 204–219.

Cox, R. H. (2001). The Social Construction of an Imperative: Why Welfare Reform Happened in Denmark and the Netherlands but Not in Germany. *World Politics*, 53(3), 463–498.

Cvetkovich, G., & Earle, T. C. (1992). Environmental Hazards and the Public. *Journal of Social Issues*, 48(4), 1–20.

Deutsch, M. (1975). Equity, Equality, and Need: What Determines Which Value will be Used as the Basis of Distributive Justice? *Journal of Social Issues*, 31(3), 137–149.

Dimitrakopoulos, D. G. (2005). Norms, Interests and Institutional Change. *Political Studies*, 53(4), 676–693.

Dryzek, J. S. (2005). *The Politics of the Earth: Environmental Discourses*. Oxford University Press.

Dudley, G. (1999). Competing Advocacy Coalitions and the Process of "Frame Reflection": A Longitudinal Analysis of EU Steel Policy. *Journal of European Public Policy*, 6(2), 225–248.

Fischer, F. (1993). Citizen Participation and the Democratization of Policy Expertise: From Theoretical Inquiry to Practical Cases. *Policy Sciences*, 26(3), 165–187.

Fischer, F. (2000). *Citizens, Experts, and the Environment: The Politics of Local Knowledge*. Durham, NC: Duke University Press.

Florini, A. (1996). The Evolution of International Norms. *International Studies Quarterly*, 40(3), 363–389.

Futrell, R. (2003). Framing Processes, Cognitive Liberation, and NIMBY Protest in the US Chemical-Weapons Disposal Conflict. *Sociological Inquiry*, 73(3), 359–386.

Gerrard, M. B. (1996). *Whose Backyard, Whose Risk: Fear and Fairness in Toxic and Nuclear Waste Siting*. Cambridge: MIT Press.

Glasgow, J. (2005). Not in Anybody's Backyard: The Non-Distributive Problem with Environmental Justice. *Buffalo Environmental Law Journal*, 13, 70–123.

Goldstein, J., & Keohane, R. O. (eds.). (1993). *Ideas and Foreign Policy: Beliefs, Institutions, and Political Change*. Cornell University Press.

Gourevitch, P. A. (1989). Keynesian Politics: The Political Sources of Economic Policy Choices. In P. A. Hall (ed.), *The Political Power of Economic Ideas Keynesianism across Nations* (pp. 87–106). Princeton, NJ: Princeton University Press.

Haas, P. M. (1992). Introduction: Epistemic Communities and International Policy Coordination. *International Organization*, 46(1), 1–35.

Hajer, M. A. (1997). *The Politics of Environmental Discourse: Ecological Modernization and the Policy Process*. New York: Oxford University Press.

Hajer, M. A., & Versteeg, W. (2005). A Decade of Discourse Analysis of Environmental Politics: Achievements, Challenges, Perspectives. *Journal of Environmental Policy & Planning*, 7(3), 175–184.

Hall, P. A. (1989). *The Political Power of Economic Ideas: Keynesianism across Nations*. Princeton, NJ: Princeton University Press.

Hall, P. A. (1993). Policy Paradigms, Social Learning, and the State: The Case of Economic Policymaking in Britain. *Comparative Politics*, 25(3), 275–296.

Hall, P. A. (1996). Political Science and the Three New Institutionalisms. *Political Studies*, 44, 936–957.

Hansen, R., & King, D. (2001). Eugenic Ideas, Political Interests, and Policy Variance: Immigration and Sterilization Policy in Britain and the U.S. *World Politics*, 53, 237–263.

Hay, C. (2010). Ideas and the Construction of Interests. In D. Béland & R. H. Cox (eds.), *Ideas and Politics in Social Science Research*. New York: Oxford University Press.

Heclo, H. (1974). *Modern Social Policies in Britain and Sweden*. New Heaven: Yale University Press.

Heiman, M. K. (1990). From "Not in My Backyard!" to "Not in Anybody's Backyard!". *Journal of the American Planning Association*, 56(3), 359–362.

Heiman, M. K. (1996). Race, Waste, and Class: New Perspectives on Environmental Justice. *Antipode*, 28(2), 111–121.

Hofmann, J. (1995). Implicit Theories in Policy Discourse: An Inquiry into the Interpretations of Reality in German Technology Policy. *Policy Sciences*, 28, 127–148.

Howorth, J. (2004). Discourse, Ideas, and Epistemic Communities in European Security and Defence Policy. *West European Politics*, 27(2), 211–234.

Hubbard, P. (2005). Accommodating Otherness: Anti-asylum Centre Protest and the Maintenance of White Privilege. *Transactions of the Institute of British*, 52–65.

Hunold, C., & Young, I. M. (1998). Justice, Democracy, and Hazardous Siting. *Political Studies*, 46, 82–95.

Hunter, S., & Leyden, K. M. (1995). Beyond NIMBY: Explaining Opposition to Hazardous Waste Facilities. *Policy Studies Journal*, 23(4), 601–619.

Inhaber, H. (1998). *Slaying the Nimby Dragon*. New Brunswick: Transaction Publishers.

Kang, M., & Jang, J. (2013). NIMBY or NIABY? Who Defines a Policy Problem and Why: Analysis of Framing in Radioactive Waste Disposal Facility Placement in South Korea. *Asia Pacific Viewpoint*, 54(1), 49–60.

King, A. (1973). Ideas, Institutions and the Policies of Governments: A Comparative Analysis: Parts I and II. *British Journal of Political Science*, 3(3), 291–313.

King, D. S., & Smith, R. M. (2014). "Without Regard to Race": Critical Ideational Development in Modern American Politics. *The Journal of Politics*, 76(4), 958–971.

Kraft, M. E., & Clary, B. B. (1991). Citizen Participation and the Nimby Syndrome: Public Response to Radioactive Waste Disposal. *Political Research Quarterly*, 44(2), 299–328.

Kunreuther, H. (1986). A Sealed-bid Auction Mechanism for Siting Noxious Facilities. *The American Economic Review*, 76(2), 295–299.

Kunreuther, H. (1996). Voluntary Procedures for Siting Noxious Facilities: Lotteries, Auctions, and Benefit Sharing. In D. Munton (ed.), *Hazardous Waste Siting and Democratic Choice*. Washington D.C.: Georgetown University Press.

Laffey, M., & Weldes, J. (1997). Beyond Belief: Ideas and Symbolic Technologies in the Study of International Relations. *European Journal of International Relations*, 3(2), 193–237.

Lake, R. W. (1993). Planners' Alchemy Transforming NIMBY to YIMBY: Rethinking NIMBY. *Journal of the American Planning Association*, 59(1), 87–93.

Lake, R. W. (1996). Volunteers, NIMBYs, and Environmental Justice: Dilemmas of Democratic Practice. *Antipode*, 28(2), 160–174.

Legro, J. W. (2000). The Transformation of Policy Ideas. *American Journal of Political Science*, 44(3), 419–432.

Lesbirel, S. H. (2011a). Facility Siting: The Theory-practice Nexus. In T. Fung, S. H. Lesbirel, & K. Lam (eds.), *Facility Siting in the Asia-pacific: Perspectives on Knowledge Production and Application* (pp. 7–32). Hong Kong: The Chinese University Press.

Lesbirel, S. H. (2011b). Project Siting and the Concept of Community. *Environmental Politics*, 20(6), 826–842.

Lieberman, R. C. (2002). Ideas, Institutions, and Political Order: Explaining Political Change. *American Political Science Review*, 96(4), 697–712.

Lieberman, R. C. (2010). Ideas and Institutions in Race Politics. In D. Béland & R. H. Cox (eds.), *Ideas and Politics in Social Science Research*. New York: Oxford University Press.

Litfin, K. T. (1994). *Ozone Discourse: Science and Politics in Global Environmental Cooperation*. New York: Columbia University Press.

Marshall, C. (2000). Policy Discourse Analysis: Negotiating Gender Equity. *Journal of Education Policy*, 15(2), 125–156.

Mazmanian, D. A., & Morell, D. (1992). *Beyond Superfailure: America's Toxics Policy for the 1990s*. Boston, MA: Westview Press.

McAvoy, G. E. (1999). *Controlling Technocracy: Citizen Rationality and the Nimby Syndrome*. Washington D.C.: Georgetown University Press.

Mcclymont, K., & O'hare, P. (2008). "We're not NIMBYs!" Contrasting Local Protest Groups with Idealised Conceptions of Sustainable Communities. *Local Environment*, 13(4), 321–335.

Mehta, J. (2010). The Varied Roles of Ideas in Politics from "Whether" to "How." In D. Béland & R. H. Cox (eds.), *Ideas and Politics in Social Science Research*. New York: Oxford University Press.

Merriam, S. B. (1988). *Case Study Research in Education: A Qualitative Approach*. San Francisco, CA: Jossey-Bass.

Mohai, P., Pellow, D., & Roberts, J. T. (2009). Environmental Justice. *Annual Review of Environment and Resources*, 34, 405–430.

Moon, D. S. (2013). "Tissue on the Bones": Towards the Development of a Post-Structuralist Institutionalism. *Politics*, 33(2), 112–123.

Morell, D. (1984). Siting and the Politics of Equity. *Hazardous Waste*, 1(4), 555–571.

Moschella, M. (2015). The Institutional Roots of Incremental Ideational Change: The IMF and Capital Controls after the Global Financial Crisis. *The British Journal of Politics & International Relations*, 17(3), 442–460.

Munton, D. (ed.). (1996). *Hazardous Waste Siting and Democratic Choice*. Washington, DC: Georgetown University Press.

Nakazawa, T. (2016). Politics of Distributive Justice in the Siting of Waste Disposal Facilities: The Case of Tokyo. *Environmental Politics*, 25(3), 513–534.

Nakazawa, T. (2017). What is against an Idea of Distributive Justice? Local Responses to In-Ward Waste Disposal in Tokyo. *Environmental Sociology*, 3(3), 213–225.

Oberholzer-Gee, F., Bohnet, I., & Frey, B. S. (1997). Fairness and Competence in Democratic Decisions. *Public Choice*, 91, 89–105.

O'Hare, M. (1977). "Not on My Block You Don't": Facilities Siting and the Strategic Importance of Compensation. *Public Policy*, 25(4), 407–458.

Padamsee, T. J. (2009). Culture in Connection: Re-Contextualizing Ideational Processes in the Analysis of Policy Development. *Social Politics: International Studies in Gender, State & Society*, 16(4), 413–445.

Panizza, F., & Miorelli, R. (2013). Taking Discourse Seriously: Discursive Institutionalism and Post-structuralist Discourse Theory. *Political Studies*, 61(2), 301–318.

Parsons, C. (2000). Domestic Interests, Ideas and Integration: Lessons from the French Case. *Journal of Common Market Studies*, 38(1), 45–70.

Peters, B. G., Pierre, J., & King, D. S. (2005). The Politics of Path Dependency: Political Conflict in Historical Institutionalism. *The Journal of Politics*, 67(4), 1275–1300.

Portney, K. E. (1985). The Potential of the Theory of Compensation for Mitigating Public Opposition to Hazardous Waste Treatment Facility Siting: Some Evidence from Five Massachusetts Communities. *Policy Studies Journal*, 14(1), 81–89.

Portney, K. E. (1991). *Siting Hazardous Waste Treatment Facilities: The Nimby Syndrome*. New York: Auburn House.

Rabe, B. G. (1994). *Beyond Nimby: Hazardous Waste Siting in Canada and the United States*. Washington, DC: Brookings Institution Press.

Renn, O. (2006). Participatory Processes for Designing Environmental Policies. *Land Use Policy*, 23(1), 34–43.

Renn, O., Webler, T., & Kastenholz, H. (1996). Procedural and Substantive Fairness in Landfill Siting: A Swiss Case Study. *Risk*, 7, 145–168.

Rihoux, B. & Ragin, C. C. (eds.). (2009). *Configurational Comparative Methods: Qualitative Comparative Analysis (QCA) and Related Techniques*. Thousand Oaks: Sage.

Rose, J. B. (1993). A Critical Assessment of New York City's Fair Share Criteria. *Journal of the American Planning Association*, 59(1), 97–100.

Sabatier, P. A. (1988). An Advocacy Coalition Framework of Policy Change and the Role of Policy-oriented Learning therein. *Policy Sciences*, 21(2–3), 129–168.

Schmidt, V. A. (2002a). Does Discourse Matter in the Politics of Welfare State Adjustment? *Comparative Political Studies*, 35(2), 168–193.

Schmidt, V. A. (2002b). *The Futures of European Capitalism*. New York: Oxford University Press.

Schmidt, V. A. (2003). How, Where and When does Discourse Matter in Small States' Welfare State Adjustment? *New Political Economy*, 8(1), 127–146.

Schmidt, V. A. (2008). Discursive Institutionalism: The Explanatory Power of Ideas and Discourse. *Annual Review of Political Science*, 11(1), 303–326.

Schmidt, V. A. (2010a). Reconciling Ideas and Institutions through Discursive Institutionalism. In D. Béland & R. H. Cox (eds.), *Ideas and Politics in Social Science Research* (pp. 25–42). New York: Oxford University Press.

Schmidt, V. A. (2010b). Taking Ideas and Discourse Seriously: Explaining Change through Discursive Institutionalism as the Fourth "New Institutionalism." *European Political Science Review*, 2(1), 1–25.

Schmidt, V. A., & Radaelli, C. M. (2004). Policy Change and Discourse in Europe: Conceptual and Methodological Issues. *West European Politics*, 27(2), 183–210.

Shemtov, R. (2003). Social Networks and Sustained Activism in Local NIMBY Campaigns. *Sociological Forum*, 18(2), 215–244.

Skogstad, G. (1998). Ideas, Paradigms and Institutions: Agricultural Exceptionalism in the European Union and the United States. *Governance*, 11(4), 463–490.

Smith, A., & Kern, F. (2009). The Transitions Storyline in Dutch Environmental Policy. *Environmental Politics*, 18(1), 78–98.

Smith, K. E. (2013). The Politics of Ideas: The Complex Interplay of Health Inequalities Research and Policy. *Science and Public Policy*, 41(5), 561–574.

Smith, N., Mitton, C., Davidson, A., & Williams, I. (2014). A Politics of Priority Setting: Ideas, Interests and Institutions in Healthcare Resource Allocation. *Public Policy and Administration*, 29(4), 331–347.

Snow, D. A., & Benford, R. D. (1988). Ideology, Frame Resonance, and Participant Mobilization. In B. Klandermans, H. Kriesi, & S. G. Tarrow (eds.), *From Structure to Action: Comparing Social Movement Research across Cultures* (Vol. 1, pp. 197–217). Greenwich, CT: JAI Press.

Snow, D. A., & Benford, R. D. (2000). Clarifying the Relationship between Framing and Ideology in the Study of Social Movements: A Comment on Oliver and Johnston. *Mobilization*, 5(1), 55–60.

Snow, D. A., Rochford Jr, E. B., Worden, S. K., & Benford, R. D. (1986). Frame Alignment Processes, Micromobilization, and Movement Participation. *American Sociological Review*, 51(4), 464–481.

Stake, R. E. (1995). *The Art of Case Study Research*. London: Sage Publications.

Stone, D. (1989). Causal Stories and the Formation of Policy Agendas. *Political Science Quarterly*, 104(2), 281–300.

Stone, D. (1993). The Struggle for the Soul of Health Insurance. *Journal of Health Politics, Policy and Law*, 18(2), 287–317.

Stone, D. (2001). *Policy Paradox: The Art of Political Decision Making, Revised Edition*. New York: W.W.Norton & Company.

Surel, Y. (2000). The Role of Cognitive and Normative Frames in Policy-making. *Journal of European Public Policy*, 7(4), 495–512.

Takahashi, L. M. (1997a). Information and Attitudes toward Mental Health Care Facilities: Implications for Addressing the NIMBY Syndrome. *Journal of Planning Education and Research*, 17(2), 119–130.

Takahashi, L. M. (1997b). The Socio-Spatial Stigmatization of Homelessness and HIV/AIDS: Toward an Explanation of the NIMBY Syndrome. *Social Science & Medicine*, 45(6), 903–914.

Takahashi, L. M., & Dear, M. J. (1997). The Changing Dynamics of Community Opposition to Human Service Facilities. *Journal of the American Planning Association*, 63(1), 79–93.

Tannenwald, N. (2005). Ideas and Explanation: Advancing the Theoretical Agenda. *Journal of Cold War Studies*, 7(2), 13–42.

Teravainen, T. (2010). Political Opportunities and Storylines in Finnish Climate Policy Negotiations. *Environmental Politics*, 19(2), 196–216.

Valletta, W. (1993). Siting Public Facilities on a Fair Share Basis in New York City. *The Urban Lawyer*, 25(1), 1–20.

Van Dijk, T. A. (1992). Discourse and the Denial of Racism. *Discourse & Society*, 3(1), 87–118.

Walsh, J. I. (2000). When Do Ideas Matter?: Explaining the Successes and Failures of Thatcherite Ideas. *Comparative Political Studies*, 33(4), 483–516.

Watson, M., & Bulkeley, H. (2005). Just Waste? Municipal Waste Management and the Politics of Environmental Justice. *Local Environment*, 10(4), 411–426.

Weisberg, B. (1993). One City's Approach to NIMBY: How New York City Developed a Fair Share Siting Process. *Journal of the American Planning Association*, 59(1), 93–97.

Wendt, A. (1999). *Social Theory of International Politics*. New York: Cambridge University Press.

White, L. A. (2002). Ideas and the Welfare State: Explaining Child Care Policy Development in Canada and the United States. *Comparative Political Studies*, 35(6), 713–743.

Whitehead, A., & Tsikata, D. (2003). Policy Discourses on Women's Land Rights in Sub-Saharan Africa: The Implications of the Re-turn to the Customary. *Journal of Agrarian Change*, 3(1–2), 67–112.

Wilton, R. D. (2002). Colouring Special Needs: Locating Whiteness in NIMBY Conflicts. *Social & Cultural Geography*, 3(3), 303–321.

Yin, R. K. (2009). *Case Study Research: Design and Methods*. London: Sage Publications.

3 Tokyo garbage war and rising influence of IWWD

In the early 1970s, In-Ward Waste Disposal (IWWD) became influential in siting policies. Back then, the waste management in Tokyo was on the verge of falling apart due to the skyrocketing amount of waste and the insufficient waste disposal capacity. The governor of Tokyo declared a "Garbage War" in 1971 and developed a campaign to overcome the garbage crisis. IWWD was adopted by the Tokyo Metropolitan Government (TMG) as a central principle of its campaign. However, the influence of IWWD was limited to siting incinerators in every ward; the responsibility of each ward for disposing of its own waste was almost ignored. Why was this idea of distributive justice adopted as a central principle in siting policies? Why was the concept of self-responsibility not influential? Focusing on the four variables introduced in the previous chapter and the interaction between them, this chapter will answer these questions.

Rising influence of IWWD in the early 1970s

Beyond the 1939 siting scheme

The impact of IWWD first needs to be juxtaposed against the scheme on which the siting of waste disposal facilities had been grounded since 1939. The first organised programme for incinerator siting was laid out in the Waste Disposal Plan[1] in 1939. The plan intended to site small incinerators evenly dispersed at the outskirts, while garbage in the central area was to be disposed of in the coastal area in a concentrated way. Figure 3.1 shows the nine candidate sites for incinerators in the 1939 plan. Although this siting plan was interrupted by World War II, it became the archetype of siting policies in the post-war period.

Based on this siting scheme, waste disposal facilities had been located in suburban areas and coastal areas; the central part of the 23 wards had been exempted from siting. The Metropolitan Construction Five Years Plan[2] in 1952 planned to site nine incinerators at the outskirts of the 23 wards, although it ended up with only one new incinerator successfully sited in addition to the three incinerators already put in place (Tōkyō-to Seisō Kyoku, 2000). This plan was succeeded by the Ten Years Incinerators Construction Plan[3] in 1957, which projected new incinerators to be sited at the surrounding area while garbage from the central part

Figure 3.1   Candidate sites for incinerators in the Waste Disposal Plan in 1939.

Adapted from the "Incinerator Plan Map" in Jinkai Shori Keikaku (Toshi Keikaku Tōkyō Chihō Iinkai, 1939a: 9).

was to be dumped mostly in coastal areas. The Tokyo Long-term Plan[4] in 1963 also aimed at siting eight incinerators in suburban areas and two large ones in coastal areas.[5] As a result, there were 12 incinerators by the time the first garbage war broke out, including two incinerators under reconstruction (Tōkyō-to Seisō Kyoku, 1971).[6] Figure 3.2 shows the location of incinerators and landfills in 1971. Garbage in the 23 wards had been disposed of by incinerators dispersed in the surrounding area, and by large incinerators and landfills in the coastal area. No waste disposal facilities were located in the central part of the 23 wards.

IWWD aimed to change this old siting scheme, requiring that incinerators be sited in every ward; the wards in the centre were no longer exempt. Rather, this idea of distributive justice earmarked the central wards for incinerators for the first time. IWWD was referred to as a significant principle in the Basic Principle on Waste Disposal[7] as a supplement to the Tokyo Mid-Term Plan 1971.[8] It stated that, based on IWWD, 13 new incinerators were to be constructed by 1975 in 13 wards[9] where there was no incinerator or one but not enough capacity to take care of its own waste. Back then, out of the 23 wards, 11 wards in the central part were left without incinerators or any plan to site them, and the incinerators at two wards, Ota and Arakawa, were considered too small to dispose of their own waste.[10]

This policy, known as One Ward One Incinerator (OWOI), was succeeded by the Tokyo Mid-Term Plan 1972.[11] While the 13 incinerators plan was treated as

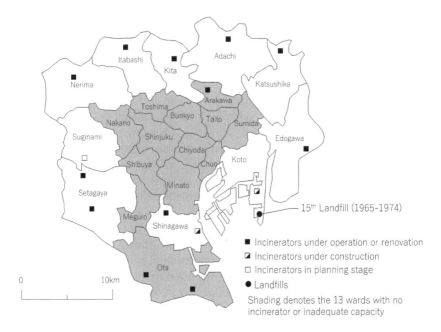

Figure 3.2   Location of incinerators and landfills in 1971.

Adapted from a map of Tokyo by CraftMap (URL: http://www.craftmap.box-i.net/) based on Tōkyō-to Kikaku Chōsei Shitsu (1972) and Tōkyō-to Seisō Kyoku (1971).

just a promise of making the utmost effort in the 1971 plan and written only in the supplement, the 1972 plan incorporated the idea more substantively. The budget for the incinerator projects was almost doubled to 97.787 billion yen from 49.819 billion yen in the 1971 plan. Thus, IWWD became influential in the siting polices of this period, changing the old siting scheme which had exempted the central part of the area from the siting of incinerators.

Responsibility of each ward and the autonomy expansion movement

On the other hand, the idea of the self-responsibility of each ward in waste disposal was hardly influential in the early 1970s. This institutional requirement of IWWD needs to be understood within the political context of the 23 wards. The 23 wards had been politically and financially less autonomous than normal municipalities under the Special Wards System. They developed a political campaign for more autonomy and the devolution of waste management was one of the most controversial issues in this political movement. IWWD, which required the devolution of the responsibility in waste disposal from the TMG to each ward, was understood in this political context.

In the 23 wards, the TMG, a regional government, controlled a large number of authorities and public services, which normally belonged to and were administrated by local municipalities.[12] After World War II, the 23 wards started as basic local municipalities,[13] as a result of the local autonomy system reform led by the General Headquarters of the Allied Powers. However, their autonomy was limited; the authority over personnel affairs, taxation and many local public services was held in the hands of the TMG. Furthermore, in the amendment of the Local Autonomy Act in 1952, the 23 wards lost the status as basic local municipalities as well as their public elections for ward mayor.[14]

This limited status of the 23 wards as local municipalities drove them to develop a political movement for more autonomy. They demanded the restoration of the public elections and the devolution of authorities which normally belonged to local municipalities. This political movement resulted in the amendments of the Local Autonomy Act in 1964 and in 1974; the movement won the devolution of some administrative services to each ward, the public elections for ward mayor, the authority over personnel affairs and so forth.

The devolution of waste management had been one of the biggest concerns in this autonomy expansion movement. Waste management has been regarded as a typical service provided by local municipalities. The Waste Cleaning Act[15] enacted in 1900 clarified that local governments were responsible for cleaning up garbage, maintaining sanitation, and disposing of collected waste.[16] Although the law did not prohibit delegating the service to private contractors, domestic waste disposal has become a responsibility of local governments since then.[17] Nonetheless, in the 23 wards of Tokyo, the TMG as the regional government had performed domestic waste disposal.[18] The 23 wards had demanded the devolution of waste management to each ward in its autonomy expansion movement. Although the devolution of waste management was written in the amendment of the Local Autonomy Act in 1964, it was suspended by the supplementary provision. The authority/responsibility for waste management remained in the hands of the TMG.

While the devolution of waste management had been negotiated, the discussions had been mostly limited to collection and transportation; devolving waste "disposal" (i.e. incineration and landfilling) to each ward had hardly been discussed. In the autonomy expansion movement, the 23 wards claimed that each ward should be responsible for and perform collection and transportation of its own waste. On the other hand, it had been thought that disposal was better performed regionally by the TMG, rather than done locally by each ward, given the uneven distribution of waste disposal facilities.

The impact of IWWD needs to be understood in this political context. The idea that a ward should be institutionally responsible for the disposal of its own waste was associated with the limited status of the 23 wards as local municipalities and the political movement for more autonomy. While the devolution of waste management services from the TMG to each ward had been one of the biggest issues in the autonomy expansion movement, the argument had been limited to garbage collection and transportation. IWWD was meant to further the devolution of waste management so that it included not only collection and transportation, but also disposal of its own waste.

However, IWWD was limited in its influence on the devolution policy during this period. Although the devolution of waste management was at issue, it was still limited to collection and transportation. Several reports from governmental advisory committees set by the TMG or the central government in the early 1970s[19] referred to the devolution of waste management to each ward. While most of them recommended the devolution of collection and transportation, waste disposal was not even on the agenda. The only exception was the Tokyo Waste Problems Special Advisory Committee's Report in 1972 (Tōkyō-to Gomi Taisaku Semmon Iinkai, 1972b). The report recommended that not only collection and transportation, but also incineration be devolved to each ward grounded on IWWD, and that necessary facilities be constructed so that each ward could perform waste management self-sufficiently. The report condemned the lack of authority/responsibility of the wards in waste management as a cause of the garbage crisis and stated that the wards should be more responsible for waste management as self-governing local municipalities.[20]

Nonetheless, this recommendation was not reflected in government policies; the argument that each ward should be responsible for waste disposal was almost ignored. Even the devolution of collection and transportation was not realised. Although the amendment of the Local Autonomy Act in 1974 repealed the supplementary provision which had suspended the devolution of collection and transportation since 1964, the amendment of the Waste Management and Public Cleaning Act[21] in the same year shelved the devolution until a day provided later by law. It was not until the 1990s that this institutional requirement of IWWD became influential.

Thus, IWWD became influential but its influence was limited to the idea of siting incinerators in every ward. While the devolution of garbage collection and transportation failed, the devolution of waste disposal was not even on the agenda. The rest of this chapter explains why this was the case. The next section looks into views and actions of Koto ward, the original claimant of this idea of distributive justice.

Koto ward's campaign against garbage pollution

Accumulating burden on Koto ward

IWWD started with a protest from one of the 23 wards against garbage pollution and its pursuit for distributive justice in waste disposal. Koto ward, located in the eastern coastal area of Tokyo, had long suffered the environmental degradation and pollution caused by the disproportionate burden of waste disposal. Figure 3.3 shows the location of the coastal landfills after the World War II.

At the time when the first garbage war broke out in 1971, around 65% of 13,888 tonnes/day of the municipal waste generated in the 23 wards of Tokyo was dumped in landfills in Tokyo Bay next to Koto ward (Tōkyō-to Seisō Kyoku, 1971).[22] A huge amount of food scraps in the dumping sites had caused bad smells and outbreaks of flies and rats. The Dream Island incident in 1965 was symbolic

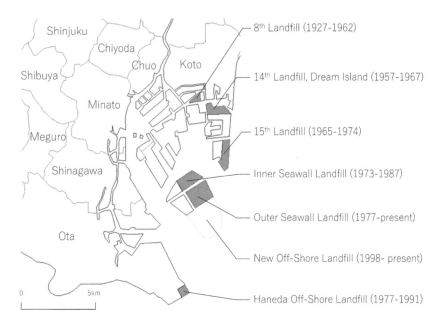

Figure 3.3   Location of the coastal landfills.

Adapted from a map of Tokyo by CraftMap (URL: http://www.craftmap.box-i.net/) based on Tōkyō-to Seisō Kyoku (2000: 648).

of the garbage pollution. A large cloud of flies bred in the food scraps piled up in the 14th landfill site, known as the Dream Island, and stormed Koto ward, which then required the Self Defence Force to burn away the garbage with flame guns (Kōtō-ku, 1965a). Furthermore, a huge number of garbage trucks caused major pollution and inconvenience to the local population. More than 5,000 garbage trucks drove through the ward every day, which made a waiting line as long as 2 km (Kōtō-ku ,1974; Tōkyō-to Seisō Kyoku, 1971). The people in the ward suffered serious traffic congestion, traffic accidents, air pollution, and spillover of dirty water from the trucks.

The burden of waste disposal on Koto ward originated during the Edo era and had become gradually concentrated in the area after World War II. In 1655, the Edo Shogunate ordered that garbage should be transported and dumped into Eitaiura, which is now a part of Koto ward, since dumping into the rivers and waterways obstructed water transportation while land reclaiming was needed as the population grew (Ito, 1982; Tōkyō-to Seisō Kyoku, 2000). Since then, waste had been dumped in landfill sites in the area of Koto.[23] Although inland landfills in Tokyo and neighbouring cities had accepted around 60% of garbage dumping until the early 1950s, the rapid urbanisation and heightened awareness of pollution among citizens made it difficult to secure sites for dumping in the inner area (Tōkyō-to Seisō Kyoku, 1952, 1960, 2000). The Metropolitan Construction Plan[24] in 1958 schemed to close all of the inland dumping sites by 1966 and to expand incineration capacity instead. However, the incineration capacity had not been sufficiently increased due to the declining performance of the old incinerators, the difficulty of finding new sites, and persistent local opposition (Tōkyō-to Seisō Kyoku, 1971).

The rapid economic growth compounded the accumulation of the burden on Koto as it brought about the rapid increase of waste generation. Japan enjoyed rapid economic growth from 1954 to 1973. The Japanese economy grew annually by nearly 10% on average during this period. Japan, which was defeated and devastated in World War II, already recovered the same economic level as before the war by 1955. The Economic White Paper in 1956 stated that "we no longer live in the post-war era" (Keizai Kikaku Chō, 1956). The Income Doubling Plan[25] announced in 1960 had achieved its goal in 1967, three years earlier than planned. This Japanese post-war economic miracle pushed the country to become the second largest economy by 1968, overtaking West Germany.

The amount of waste generated had also soared at a tremendous pace under this economic growth, combined with the advent of a mass-production and mass-consumption society. Increased income led to more and more consumption, which resulted in a growing amount of waste. Waste generation in Tokyo, which was only around 110,000 tonnes/year in 1947, jumped to more than one million tonnes/year by 1960. The amount continued to increase by around 12% annually from 1960 and reached three million tonnes/year in 1970 and 3.9 million tonnes/year in 1972 (Tōkyō-to Seisō Kyoku, 2000).

Thus, by the beginning of the 1970s, the depletion of inland dumping sites, the delay of incinerator construction, and the sharp increase of waste production under the rapid economic growth resulted in the massive influx of garbage into the landfills next to Koto ward. The shift from water to land transportation of garbage during the late 1960s, due to the construction of highways and levees, also escalated the traffic through Koto ward to the coastal landfills (Ishii, 2006; Tōkyō-to Seisō Kyoku, 2000). Koto ward had repeatedly protested the garbage dumping and asked the TMG to alleviate the environmental pollution since the Dream Island incident in 1965. However, to the contrary, the TMG proposed to extend the use of the 15th landfill in 1971 to cope with the rapid increase of waste.

That was when Koto ward declared IWWD and demanded the other wards to accept some of the burdens of waste disposal. The Koto ward council announced its opposition to the continued dumping of food scraps at the 15th landfill with "grave determination".[26] The ward sent open letters to the TMG and the other wards and demanded them to accept IWWD as a principle of waste disposal. The rise of this idea of distributive justice was triggered with this claim by Koto ward in its protest to the disproportionate burden on the ward.

Koto's strategy and IWWD

IWWD, a notion of the self-sufficient disposal, was conceived by Koto ward,[27] so that it would fit with Koto's goal and the strategy to achieve it.[28] Koto's goal was to lessen the environmental pollution caused by the garbage dumping. As the ward saw the shortage of incineration capacity as the major cause of the accumulating burden, it advocated this idea to facilitate the construction of incinerators by rousing the sense of responsibility of the other wards and getting them to be engaged with waste disposal. The two requirements of IWWD, siting incinerators in every ward and the self-responsibility of each ward for waste disposal, were derived from this strategy by Koto ward.

IWWD was tightly connected with incineration, because, by claiming IWWD, Koto ward aimed to make a shift from landfill-dependent disposal to incineration-centred disposal.[29] In theory, IWWD could have argued for all kinds of facilities necessary for waste management. Completely self-sufficient waste disposal needed not just incinerators but also landfills as well as other types of waste treatment facilities. Even if garbage for incineration were all incinerated within a ward, landfills would still be necessary to deal with the residual ashes and waste not to be incinerated. It is true that, in the open letter to the TMG, Koto demanded that all waste should be disposed of within a ward regardless of whether the waste was for incineration or not (Kōtō-ku-gikai, 1971a). However, in the letters to the other wards, it was apparent that Koto focused on incinerators, demanding cooperation for the siting of incinerators (Kōtō-ku-gikai, 1971b). In fact, IWWD was almost exclusively associated with incineration in the ward's campaign for fairness in burden distribution among the 23 wards.

This was because the shortage of incineration capacity was regarded as the major cause of the disproportionate burden and the environmental degradation in the ward. Anti-kitchen garbage dumping had been a central issue in its campaign against garbage pollution since the 1960s. Expanding incineration capacity and ending the dumping of un-incinerated food scraps were seen as the fundamental way to reduce the pollution in the ward. When the 15th dumping site was proposed in 1964, to persuade Koto ward, the TMG promised that a sufficient number of incinerators would be constructed to dispose of all waste for incineration and that kitchen garbage dumping would be stopped by 1970 (Tōkyō-to Seisō Kyoku, 1964). However, this promise was not realised by 1970 (Tōkyō-to, 1969; Tōkyō-to Kikaku Chōsei Shitsu, 1971). Although the percentage of incineration in waste disposal grew from 11.3% in 1961 to 37.5% in 1970,[30] the amount of dumped waste had risen from 1.35 million tonnes/year to 2.25 million tonnes/year in ten years (Tōkyō-to Seisō Kyoku, 1971). Furthermore, some incinerator projects had fallen behind schedule. The expansion of incineration capacity could not keep up with the soaring amount of waste, which resulted in the influx of garbage into the coastal landfills. Consequently, Koto attributed the main cause of the disproportionate burden to insufficient incineration capacity.

Accordingly, IWWD aimed to facilitate the construction of incinerators. Koto intended to get the other wards involved in the garbage problem. Before Koto began to argue for IWWD, the battle over garbage dumping had been fought only between Koto and the TMG. The rest of the wards seemed, at least to Koto, indifferent to waste management issues and the sufferings that Koto had been forced to bear. Koto intended to change this conflict structure by involving the other wards that were sending garbage to Koto. Waste disposal was not just Koto's problem, argued the ward, but should be considered a problem for the 23 wards as a whole (Mainichi Shimbun, 1971a; Yomiuri Shimbun, 1971b). IWWD aimed to raise the awareness of the other wards in waste disposal as self-governing local municipalities, and make them accept some responsibility in the construction of incinerators to rectify the distributive injustice among the wards.[31]

It is noteworthy here that IWWD was not the only approach to distributive injustice which Koto claimed. The ward demanded that the TMG take immediate measures to minimise environmental pollution. Koto ward was especially concerned about garbage trucks passing through the ward as a major source of pollution. To reduce the number of garbage trucks, the TMG promised to increase sea-transportation and tranship waste from small trucks to large ones by constructing relay stations in other wards (Tōkyō-to, 1971c).

Koto ward asked for compensation as well. There had been discontent in Koto that the ward was unfairly underdeveloped[32] and Koto had made use of the siting of waste disposal facilities as an opportunity to facilitate the development of the area. For instance, when the TMG proposed to construct an incinerator at the Dream Island site (i.e. the 14th landfill) in the 1960s, the ward requested facilities such as a swimming pool, a botanical garden, a welfare facility for the elderly and the disabled, a baseball stadium, and so forth (Kōtō-ku & Kōtō-ku-gikai, 1970). By the same token, in the open letter to the TMG, Koto complained that there were few beneficial facilities in the ward in spite of the many locally unwanted facilities located there. To appease the anger of Koto ward, the TMG offered a variety of compensation packages.[33] Drawing out as much compensation as possible was another goal for Koto in its campaign for distributive justice.[34]

However, it was IWWD which played a central role in the ward's campaign for distributive justice. A long-term, fundamental remedy was sought by increasing incineration capacity, while burden minimisation and compensation were required as makeshift measures. Furthermore, IWWD provided a normative ground to justify the other approaches. The notion of self-sufficient disposal highlighted the distributive injustice among the 23 wards and brought to light the disproportionate burden that Koto ward had long suffered, which justified Koto's demand for burden reduction measures and compensation. By showing the gap between the ideal and the reality, IWWD worked as a philosophical basis to legitimise the entire campaign for distributive justice.

Koto's pressure on the TMG

Koto ward was able to make the idea of distributive justice accepted by the TMG. In its reply to the open letter from Koto (Tōkyō-to, 1971b; Tōkyō-to-gikai, 1971b), the TMG apologised to the ward for the trouble it had suffered so far, and accepted IWWD as a basic principle of waste disposal. IWWD was translated into the OWOI policy and the 13 incinerators project was announced in early 1972. This quick adoption was a result of the powerful and persistent pressure from Koto ward.

The power of Koto is attributable to two sources: the blockade of garbage to the existing landfill and the de facto veto of the siting of new ones. Koto ward threatened the TMG with the blockade of garbage trucks coming through the ward to the existing landfill. In negotiations with the TMG, Koto repeatedly referred to a "grave determination", which suggested stopping the waste from being transported through the ward (Asahi Shimbun, 1971; Kōtō-ku-gikai, 1971a, 1971c). As waste disposal was heavily dependent on the 15th landfill as the final destination, waste management would have fallen apart if waste into that landfill had been blocked.

Koto repeatedly threatened the TMG and the rest of the wards with the blockade to make them accept IWWD as well as other measures to lessen the pollution. The Koto ward council announced its intention to block waste into the landfill in November 1971, being dissatisfied with the abstract measures which the TMG proposed in the reply to the first open letter. To prevent the blockade, the TMG had to withdraw the proposal of extending the 15th landfill, show more concrete measures to reduce the burden on Koto, and promise a rapid resolution of the conflicts over incinerator siting in Suginami, Adachi, and Kasai wards, where neighbours' opposition had delayed the projects.

Actually, the blockade was implemented twice during the garbage war.[35] It targeted Suginami ward where the incinerator siting at Takaido in the ward had been delayed for a long time by local opposition.[36] Waste from Suginami was blocked at the end of 1972 and in May 1973. Koto announced the blockade, and members of the ward council checked and stopped the garbage trucks from Suginami at the entrance of the 15th landfill. The blockade lasted for one day in the first instance, but continued for three days in the second; Suginami was buried in piles of garbage.

Furthermore, Koto had political veto power in the siting of the new landfill. After the extension of the 15th landfill was cancelled, the TMG had to secure new ones as soon as possible in the face of the skyrocketing amount of waste. The process to site landfills in the sea was regulated by the Act on Reclamation of Publicly-owned Water Surface.[37] This act required hearing the opinions of concerned local governments over the projects. Even though the candidate site of the new landfill did not belong to Koto,[38] the process required the TMG to hear Koto's opinion, as garbage trucks would pass through the ward and impact its environment. While the approval of local municipalities concerned with new landfills was not a legal requirement, this necessity to hear the opinions of the affected wards worked as a de facto veto. Although autonomy of the wards had been limited compared to other municipalities, each one of the 23 wards was an independent local autonomy with its own government and council; the TMG could not ignore the opinion of a ward over issues such as locally unwanted facility siting.

This siting process provided the ward political opportunities to make its claim heard by the TMG. The TMG proposed two new landfills at the inner and outer seawall areas by urgent recommendation from the Tokyo Waste Problems Special Advisory Committee (Tōkyō-to Gomi Taisaku Semmon Iinkai, 1972a). Koto made use of this opportunity to make the TMG take IWWD more seriously. In June 1972, Koto agreed on the inner seawall landfill but on the condition that the project would be stopped if the incinerator projects did not make substantive progress[39]; the projects in Suginami, Adachi and Katsushika[40] came to an agreement by December 1972, March 1973 and August 1973 respectively, and more than half of the 13 incinerators were agreed to by June 1973. Although the inner seawall landfill came to an agreement and started operation at the end of 1973, the TMG had to continue negotiation with Koto ward over the outer seawall landfill, as the inner seawall landfill was not expected to last very long. Again, Koto ward suspended its agreement and drew out promises and conditions from the TMG to promote IWWD as well as other measures to redress the disproportionate burden.

It is worth mentioning that the rapid economic growth accounts for this power of Koto ward as well as the accumulation of the burden on the ward. What made the blockade and the political veto effective bargaining chips in the negotiation was the heavy dependence on the coastal landfill, both then and in the prospective future. Koto was powerful as it could take advantage of this over-dependence during the deepening landfill crisis. As noted earlier, the skyrocketing amount of waste resulting from rapid economic growth accelerated the depletion of inland landfills, leaving the coastal area as the only place available for dumping. The capacity of the 15th landfill was also being consumed quickly by the soaring amount of waste, thereby making the TMG desperate to construct the new landfills. The blockade and the political veto over the new landfills were all the more effective given this landfill crisis.

Tokyo Metropolitan Government's adoption

IWWD was adopted by the TMG fairly quickly. Facing the protest from Koto, the TMG declared a "Garbage War" and started a campaign for overcoming the garbage crisis. IWWD became the central idea, or the slogan, in this war against garbage.[41] The governor of Tokyo, Ryokichi Minobe, stated that the cause of this garbage crisis was in the policies which had prioritised economic development over the people's quality of life, as well as in the people's disregard of the waste problem once taken out of their sight. To solve the garbage crisis, argued the governor, it needed to raise citizens' awareness of waste issues and build waste disposal facilities necessary for their lives. The solution was sought by siting waste disposal facilities in everyone's community. IWWD was adopted as the fundamental principle to support this argument and was reflected in the siting policies.

This immediate adoption of IWWD was not just because Koto ward was powerful in the policy-making process, but also because this idea of distributive justice was congruent with basic values which had underlain the siting policies of the TMG, and the TMG recognised it as a solution to the garbage crisis.

Congruence with values underlying TMG's siting policy

IWWD was accepted by the TMG partly because this idea was congruent with two values which had underpinned the siting scheme in the 23 wards of Tokyo: the efficiency in garbage transportation and the autonomy of each community in waste disposal. Both of them would be satisfied with a large number of small incinerators evenly distributed over the 23 wards of Tokyo rather than a small number of large incinerators, thereby satisfying the requirement of IWWD. As explained earlier, the grand design of incinerator siting in the post-war period originated in the 1939 siting plan. As shown in Figure 3.1, this plan divided the new Tokyo area, which was the surrounding area newly merged into Tokyo City during the 1930s, into nine garbage collection areas and intended to site nine incinerators, one in each area, along beltways.[42]

This siting scheme aimed to revise the concentrated waste disposal with a small number of large incinerators, learning a lesson from the Fukagawa Incident. After several failed attempts since 1900, three large incinerators were constructed in Fukagawa ward, which is now a part of Koto ward. However, right after the second and the third plants had started operation in 1933, people in the ward protested against the air pollution caused by these incinerators. As waste from all of the wards were incinerated in Fukagawa, the people questioned why only Fukagawa ward had to accept the burden of incineration. A member of the Tokyo City Assembly from the ward claimed that the concentrated disposal should be shifted to dispersed disposal and that incinerators should be sited in every ward.[43]

Experiencing this incident, the 1939 siting plan criticised the concentrated incineration in Fukagawa for causing the intense pollution and the inefficiency in transportation of garbage. Accordingly, the 1939 plan was based on the lesson learnt that siting incinerators dispersedly was more desirable than a concentrated disposal, emphasising the efficiency in garbage transportation and the notion of autonomy – that one's garbage should be taken care of by oneself (Toshi Keikaku Tōkyō Chihō Iinkai, 1939b, 1939c).[44] Although the 1939 plan was interrupted by World War II, this siting scheme was inherited in the siting policies after the war as the blueprint.

However, this siting scheme had been applied only to the surrounding areas of Tokyo; the central area had been exempted from waste disposal facilities. The 1939 siting plan was intended to dispose of waste in the newly merged area of Tokyo. In 1932, 82 towns and villages were merged into Tokyo City and organised into 20 wards in addition to the existing 15 wards, which made the Great Tokyo City almost equivalent in size to the 23 wards of Tokyo today.[45] The suburb area was being rapidly urbanised as the Tokyo Great Earthquake in 1923 devastated the central area of Tokyo. This rapid urbanisation led to the increase of waste generation and made it more difficult to find new inland dumping sites there. This necessitated new incinerators for the new area, while waste from the old, central part of Tokyo was to be incinerated at the Fukagawa plants.

In total, the entire siting scheme was a compromised combination of the dispersed disposal in the suburban areas and the concentrated disposal in the central area.[46] The siting policies after the war had also been based on the same scheme which intended to site small incinerators evenly dispersed over the suburb area, while garbage in the central area was to be disposed of in the coastal area in a concentrated manner. In other words, the application of the two values, i.e. the efficiency in transportation and the autonomy of each ward, was limited to the suburb area.

IWWD was to extend these pre-existing values to the central area which had been exempted in this old 1939 siting scheme. As noted earlier in this chapter, IWWD aimed to change this old siting scheme. However, IWWD was neither replacing nor challenging these values which had underpinned the old siting scheme. Rather, this idea of distributive justice was meant to strengthen them by broadening their application to the central area. In fact, even before IWWD was proposed by Koto ward, siting incinerators in every ward had been recognised as normatively legitimate from the view point of transportation efficiency and the autonomy/responsibility of each ward,[47] although its application in the central wards was considered difficult, if not impossible, due to the high density and the limited land availability there.[48] This idea of distributive justice was appealing to the TMG because it resonated with the long-held values which had informed its previous siting policies.

Furthermore, transport efficiency and the autonomy of each ward had become more compelling by the time Koto ward argued for IWWD. Back then, the severe traffic congestion resulting from the intense urbanisation and rapid motorisation crippled waste management in Tokyo (Tōkyō-to Gomi Taisaku Semmon Iinkai, 1972b; Tōkyō-to Seisō Kyoku, 1971). The depletion of inland landfills and the delay of incinerator construction left a large amount of waste to be brought to the coastal landfills, thereby making the distance of garbage transportation much longer. The traffic congestion and the long-distance transportation hindered more frequent waste collection and left garbage on the streets in unsanitary conditions. The TMG had to increase the number of workers and trucks to cover the time loss, which pushed up the cost of transportation. In the early 1970s, the collection and transportation cost amounted to 76.4% of the total waste management cost (Tōkyō-to Seisō Kyoku, 1971). IWWD was appealing to the TMG as this idea of distributive justice would make the transportation distance as short as possible, requiring that waste should be disposed of where it was generated.[49]

Autonomy of the 23 wards was also becoming normatively more compelling given the rise of the autonomy expansion movement after World War II. As explained earlier in this chapter, the limited status of the 23 wards under the Special Wards System led to the political movement for more autonomy and independence. The idea of self-sufficient disposal fitted with the notion that a ward should take care of its own waste. In particular, Minobe had been supportive of this autonomy expansion movement. Actually, it was under the Minobe administration that the public elections for ward mayor was restored and the authority over personnel affairs was handed over to the wards. As each ward was becoming more autonomous and independent, the idea that each ward should be responsible in waste disposal became all the more convincing.

IWWD as a policy solution to the garbage crisis

Furthermore, siting incinerators in every ward was recognised as a solution to existing policy problems, providing a clear roadmap out of the garbage crisis. This recognition of IWWD as a policy solution needs to be understood in relation to the All Waste Incineration (AWI) policy and the pressure to expand incineration capacity.

IWWD was strongly associated with AWI which had been one of the foremost policy goals for the TMG in waste management since the early 1960s. AWI was the policy aimed at incinerating all waste that should be incinerated.[50] Until 1960, the TMG's disposal scheme had been a mixture of dumping and incineration; waste in the central area was to be brought into the coastal landfills and that of the suburbs to incinerators at the outskirts of the 23 wards (Tōkyō-to Seisō Kyoku, 1960; Tōkyō-to-gikai, 1959c, 1960b). While the Waste Management Future Plan[51] in 1960 schemed to enlarge the percentage of incineration from 11.1% in 1959 to 49.3% by 1970, 48.8% was still to be sent into landfills in this plan.[52] AWI policy was introduced in 1961 at the Tokyo Metropolitan Assembly and written into the Tokyo Long-term Plan in 1963.[53] The plan aimed at achieving AWI by 1970 and to reduce the amount of waste directly sent to landfills to zero (Tōkyō-to, 1963; Tōkyō-to-gikai, 1961). Since then, AWI had been one of the primary goals for the TMG.

Behind this policy goal were the rapidly growing economy and the belief in incineration as the best technology to address waste management. As noted earlier in this chapter, the rapid economic growth brought about the rapid increase in waste production and put tremendous pressure on the waste disposal system which had been dependent on landfills. While the rapid urbanisation and heightened awareness of pollution among citizens made it more difficult to secure further dumping sites in the inner area (Tōkyō-to Seisō Kyoku, 1952, 1960, 2000), the coastal landfills were reaching their limit due to the skyrocketing amount of waste. A waste disposal system that depended heavily on landfills was no longer sustainable.

Incinerationism thus provided an interpretive frame to solve that problem: constructing more incinerators to minimise the volume of garbage brought into landfills. Incinerationism had been the policy paradigm which had dominated waste disposal policy in Japan. This paradigm was characterised by the technological belief in incineration and its exclusive focus on the "disposal" phase rather than the upper stream management such as source reduction. It dates back to the Waste Cleaning Act in 1900 which provided that incineration was the most desirable method, mainly due to the concern over sanitation as outbreaks of cholera were the most serious issue in waste management back then. Although this act encouraged incineration to the extent possible, the amendment in 1930 made incineration of waste a duty for local municipalities. Since then, incineration has been believed to be the most desirable and modernised way to deal with waste in Japan. The depletion of dumping sites under rapid economic development made enlarging incineration capacity even more urgent. The Japanese government had promoted the construction of incinerators through the Waste Disposal Facilities Construction Plans.[54] These five-year plans encouraged local municipalities to construct incinerators by providing subsidies. Given this policy of the central government, the percentage of incineration in waste disposal in Japan steadily rose from 45% in 1965 to 61% in 1970 (Yagi, 2004).[55]

In Tokyo as well, incineration had been regarded as the most desirable technology for waste disposal since the turn of the 20th century (Tōkyō-to, 1963; Tōkyō-to Seisō Kyoku, 1960, 2000; Toshi Keikaku Tōkyō Chihō Iinkai, 1939b). Although the construction of incinerators had been slow and lagged behind other big cities such as Osaka (Tōkyō-to Seisō Kyoku, 1971),[56] the depletion of dumping sites made enlarging incineration capacity an immediate concern. It was believed that expanding incineration capacity was the only way to solve the garbage problems. Thus, the rapid increase of waste resulting from the rapid economic growth and incinerationism as the policy paradigm resulted in the tremendous pressure to expand incineration capacity. It was under this pressure that AWI became the primary policy goal in waste management.

However, AWI was not achieved as planned because the quantity of garbage increased beyond TMG's estimation and because of persistent local opposition to the incinerator projects. The Tokyo Long-term Plan in 1963 estimated that the amount of waste for incineration in 1970 would be 7,970 tonnes/day, while the actual amount reached 10,490 tonnes/day (Tōkyō-to Kikaku Chōsei Shitsu, 1972). Furthermore, the incinerator siting projects met intense local opposition.[57] Although the Kita plant, which was announced in 1961, finally started operations in 1969 after intense local opposition, the Suginami plant at Takaido, which was announced in 1966, was still in the middle of the conflict when the first garbage war broke out. As a result, only 41.4% of waste for incineration was incinerated in 1970 and the achievement of AWI was postponed to 1975 (Tōkyō-to Kikaku Chōsei Shitsu, 1971). It was also estimated that the amount of waste would increase at the same rate in the future and that an increase in incineration capacity of 5,310 tonnes/day would be necessary by 1985 to achieve AWI (Tōkyō-to Kikaku Chōsei Shitsu, 1971), which amounted to 10 more incinerators with 600 tonnes/day capacity.

IWWD was therefore thought to be the solution to this garbage crisis because this idea set out a clear roadmap to achieve AWI. For the TMG, the problem was that AWI would not be achieved and waste disposal would be disrupted as incineration capacity expansion could not keep up with the rapid waste increase. But IWWD set out how AWI could be achieved: by constructing incinerators in every ward and ensuring that waste generated in a ward be incinerated in that ward. AWI became increasingly difficult to achieve with the previous siting scheme which exempted the central wards. The amount of waste was soaring far beyond the previous estimation of the incinerator siting plans. IWWD was appealing to the TMG because this idea of distributive justice was expected to facilitate the incinerators construction in the central area and help achieve AWI amidst the rapid garbage growth.[58]

In other words, the policy rationale of IWWD coincided with the interest of Koto ward: reducing the amount of waste flowing into the landfills as much as possible by increasing incineration capacity. Making waste disposal less dependent on landfills by constructing more incinerators was their common goal. The strength of IWWD stood on this agreement between the policy goal of the TMG and the interest of Koto ward.

Doubt as to its feasibility

It is noteworthy that the practicability of the OWOI policy was doubted by the TMG, thereby undermining the cognitive legitimacy of the idea. Land availability had been the biggest barrier to siting incinerators in the central area of Tokyo. The 1939 siting scheme was based on the assumption that it was impossible to construct incinerators in the central wards.[59] That was why the dispersion of incinerators had been applied only to the suburb area, and garbage in the central wards had been sent to the landfills in the coastal area. Even before IWWD was proposed, it was argued occasionally in the Tokyo Metropolitan Assembly that an incinerator should be sited in every ward. However, the TMG had been resistant to this argument because of land availability; it reflected the perception right before the first garbage war that incinerators were impossible to site in the central wards such as Chiyoda, Chuo, Shinjuku, Bunkyo, Shibuya, and Toshima (Tōkyō-to-gikai, 1971a).

As IWWD aimed to extend the siting of incinerators to the central wards, the TMG had to face this problem again. An incinerator needed spacious land that could be accessed via wide roads so that hundreds of large garbage trucks could pass through. Finding such land was extremely difficult in the 23 wards. It was all the more so in the hyper-congested central and sub-central parts of Tokyo. IWWD, which required a large number of small incinerators rather than a small number of large ones, was accepted by the TMG partly due to its perception that it was already impossible to secure vast lands for large incinerators in the 23 wards of Tokyo (Tōkyō-to-gikai, 1972b). However, even for small incinerators, it was difficult to secure sites in the central part.[60]

This was why the joint waste disposal by several neighbouring wards was allowed as an exception to IWWD. In the reply to Koto's open letter, the TMG tried to leave the option of having one incinerator shared with two or three wards (Tōkyō-to, 1971b). At first, the TMG used the term "In-Community Waste Disposal", instead of "In-Ward Waste Disposal", which meant that waste of a community should be taken care of within the community, without clarifying its geographical scale.[61] Even Koto ward left the option of joint disposal in the open letter to the other wards, asking them whether or not they would take part in a joint disposal if OWOI were difficult to achieve.

Furthermore, in response to the question on the feasibility of the 13 incinerators project, the TMG stated that this should be treated as just a promise of making the utmost effort and hence was written only in the supplement of the Tokyo Mid-term Plan 1971 (Tōkyō-to-gikai, 1972a). The 13 incinerators project was no more than an abstract roadmap without any candidate sites for them. This feasibility issue undercut the cognitive legitimacy of IWWD and thereby compromised the application of the principle in the siting policy.

Nonetheless, the influence of IWWD became even stronger in the Tokyo Mid-term Plan 1972. This was partly because of the pressure from Koto. The TMG had to show its determination to achieve IWWD in order to persuade the ward to approve the new landfill projects. In addition, although the completion was supposed to be difficult to attain, the TMG expected that IWWD would facilitate the incinerator siting and help achieve AWI. As noted above, the TMG advocated IWWD due to the prospected need for more incinerators for AWI. The TMG made use of IWWD and the pressure from Koto to facilitate the incinerator siting necessary for this policy goal.

The self-responsibility of each ward and the devolution of waste management

Union's intervention and the TMG's hesitation

On the other hand, the self-responsibility of each ward was hardly an influential idea in this period while the idea of siting incinerators in every ward became dominant. This was mainly because self-responsibility was against the interest of the Tokyo Cleaning Workers Union. The union did not oppose the siting of incinerators in every ward; but it feared the idea of facilitating the devolution of waste management to each ward.

The union rather supported the idea of constructing more incinerators. It argued that the garbage crisis occurred as a result of people's indifference to waste management and discrimination against the cleaning workers (Tōkyō Seisō Rōdōkumiai, 1981, 1999). Waste management had been disregarded and the cleaning workers had been suffering from discrimination;[62] waste and waste-related facilities had been excluded from communities. People did not care once their waste was taken out of their living environment. Poor facilities and vehicles further exacerbated the neighbours' aversion to these facilities. To resolve the garbage crisis, argued the union, it was essential to change people's perception of waste and to construct proper facilities necessary for waste disposal. To make waste-related facilities acceptable to communities, they had to be modern, sanitary and clean. Improving facilities would lead to a better working environment for the cleaning workers as well. The idea that the garbage from a community should be disposed of in the community provided the philosophical basis for these claims. IWWD, siting incinerators in every ward, was recognised not only as being cognitively legitimate as a policy, but also as serving their interests, i.e. the improvement of their work conditions.

On the other hand, the union feared that this idea would also facilitate the devolution of waste management. The union was opposed to devolving the responsibility of waste management to each ward because it was against its organisational interest. The union was afraid that their centralised organisation would have to be divided into 23 "local" entities if waste management authority was devolved. This would have weakened the power of the organisation as a whole, thereby increasing the risk of deterioration in the working conditions of union members. Even the devolution of collection and transportation was out of the question, for it was these sections which had the largest number of workers.

As IWWD was associated with the autonomy expansion of the wards, the union had to prevent this idea from facilitating the devolution of waste management.

In fact, right after Koto ward declared IWWD, the union quickly reacted and took actions to stop it from influencing the politics on devolution (Tōkyō Seisō Rōdōkumiai, 1971). The union visited high officials of the TMG, such as the head of the Bureau of Waste Management, the lieutenant governor, and the head of the Office of Planning and Coordination,[63] and asked them not to devolve waste management to the wards (Tōkyō Seisō Rōdōkumiai, 1971).

Facing pressure from the union, the Minobe administration of the TMG became less positive about devolving waste management to each ward. Responding to the concerns from the union against the devolution, the officials of the TMG promised that waste management would not be devolved (Tōkyō Seisō Rōdōkumiai, 1971). Minobe stated in the assembly that the TMG would remain in charge of waste management for the time being, when asked about the relationship of IWWD with the autonomy expansion and the governor's willingness to devolve authority.[64] Although devolving the responsibility for waste disposal to each ward was a natural consequence of IWWD, stated the governor, it was difficult and would aggravate the crisis, given the lack of and uneven distribution of waste disposal facilities. Although the report of the Tokyo Waste Problems Special Advisory Committee, which was formed to make proposals to tackle the garbage crisis, recommended that the responsibility of waste management, except for landfill disposal, should be devolved to each ward according to IWWD, the TMG tried to separate the institutional responsibility of each ward away from this idea of distributive justice and limited the argument to constructing incinerators in every ward.

This was partly because it was difficult for Minobe and the TMG to disregard the union's interest, even though Minobe was eager for autonomy expansion and devolution. Overcoming the garbage crisis was impossible without the union's cooperation and support; the TMG could not afford to fight against the workers during the deepening crisis. Moreover, the Minobe administration was one of the "Kakushin Jichitai",[65] i.e. a progressive local government led by left-wing mayors/governors with the support of the Japanese Socialist Party and/or the Japanese Communist Party. In the early half of the 1970s, there were a number of local governments which were led by socialist/communist heads while the Liberal Democratic Party had been in power at a national political level. They acquired popularity, especially in big cities,[66] by taking a strong stance against environmental pollution and improving welfare policies. Minobe, who was a scholar of Marxist economics, was one of those heads and came into office in 1967 backed by the Japanese Socialist Party and the Japanese Communist Party. Minobe, therefore, could not ignore the opinions of the union, which was one of the significant advocates for his government.[67]

Furthermore, even if the amendment of the Local Autonomy Act were submitted by the Japanese central government, the union was able to prevent devolution through the influence of the Japanese Socialist Party, which was still influential in the national politics. Although the Japanese Liberal Democratic Party had dominated national politics since 1955, the Japanese Socialist Party had been the biggest opposition with nearly a third of the seats in the House of Representatives back then. In fact, the Japanese Socialist Party was the one who opposed the devolution of collection and transportation in the amendment of the Local Autonomy Act in 1964 and won the supplementary provision, which left waste-related services in the hand of the TMG.[68]

Besides the interest and power of the union, it was recognised that the devolution would make the problem worse rather than solve it. The union argued that waste management was better performed regionally by the TMG rather than locally by each ward. The TMG also thought that devolving the responsibility was premature. Even the devolution of collection and transportation had been recognised as difficult due to the insufficient and uneven distribution of waste-related facilities. Devolving the responsibility in waste disposal was all the more impossible as less than half of the 23 wards had incinerators. As the waste management of Tokyo was at the verge of falling apart, given the rapid increase of waste generation and Koto's opposition to the new landfills, devolving the responsibility to each ward was recognised as problem causing rather than problem solving.

Dilemma of the wards

The other wards were also involved in this garbage war, as IWWD required them to accept the burden and the responsibility of waste disposal to rectify the injustice among the wards. The support of the wards was indispensable for IWWD to be successful. Facility siting could not proceed if a ward government and/or council opposed the project.[69] The responsibility in waste disposal could not be devolved without their agreement. During the garbage war, they remained rather lukewarm and ambivalent towards IWWD; they were reluctant to take institutional responsibility for incineration while generally agreeing with the siting of incinerators in every ward.

All the wards generally agreed on the idea of constructing incinerators to dispose of their own waste. In the replies to the open letter from Koto, which asked whether they would agree to IWWD and to cooperate in siting incinerators, they supported the idea in general. Some of the wards in which incinerators had already been sited or were being planned were rather understanding of Koto (Kōtō-ku-gikai, 1971b). It is true that others showed cautious attitudes to avoid having additional burdens imposed on them. For instance, Nerima ward and Shinagawa ward maintained that there had been enough incinerators for their own waste and an additional one was not acceptable (Kōtō-ku-gikai, 1971b). The Shinagawa ward council resolved to block waste from Suginami when Koto ward announced the blockade in May 1973 (Tōkyō-to Gomi Taisaku Hombu, 1973). In Nerima ward, neighbours around the existing incinerator at Shakujii protested against waste coming from other wards (Kōtō-ku-gikai, 1971b; Tōkyō-to-gikai, 1972a). However, they were supportive of IWWD in principle. Even the 13 wards without enough incinerators agreed to IWWD and promised their cooperation. In fact, when the TMG spoke to the mayors and the chairpersons of the 13 wards in February 1972, all of them promised to cooperate in finding candidate sites for incinerator siting (Asahi Shimbun, 1972a; Mainichi Shimbun, 1972; Yomiuri Shimbun, 1972b).

On the other hand, they, the 13 wards in particular, were reluctant to be institutionally responsible for waste disposal. They contended that waste disposal should be regionally managed by the TMG rather than locally by each ward although they promised to help the TMG site incinerators. Behind this partial support for IWWD, there was complicated interaction between conflicting views on distributive justice, doubts about the practicability of siting an incinerator in every ward, and the congruence with the value and the interest of the autonomy expansion movement.

Some of the 13 wards that were targeted by Koto initially showed rather negative attitudes towards IWWD. Minato, Shinjuku and Chuo wards claimed that they also suffered from other kinds of locally unwanted facilities, such as a sewage treatment plant, a human waste disposal plant and a slaughterhouse. For instance, Minato ward argued that a sewage treatment plant in the ward disposed of sewage from other wards while a slaughterhouse there provided 25% of the meat consumed in Tokyo (Kōtō-ku-gikai, 1971b). Chuo ward also complained that most of the human waste treated by a facility in the ward came from other wards (Kōtō-ku-gikai, 1971b). Shinjuku also had a sewage disposal plant which was used for other wards (Yomiuri Shimbun, 1971b). What types of locally unwanted land uses should be counted is important in assessing distributive justice. As IWWD concerned the siting of incinerators almost exclusively, this idea of distributive justice did not take into account the distribution of other types of locally unwanted facilities. Accordingly, the idea was ideationally less convincing to those who bore burdens other than incinerators.

The self-responsibility of each ward in waste disposal was not acceptable for the wards in which a site for an incinerator was hard to find. Most of the 13 wards, Chiyoda, Chuo, Minato, Shinjuku, Bunkyo, Taito, Sumida, Meguro, Nakano, and Toshima, doubted the practicability of IWWD, because they thought that appropriate lands for such use were hardly available (Kōtō-ku-gikai, 1971b; Mainichi Shimbun, 1971b, 1973; Yomiuri Shimbun, 1971a, 1971c, 1972a). Given this limited land availability, they argued that disposing of waste within each ward was impossible, and consequently waste disposal should be performed regionally by the TMG. The feasibility issue undermined the credibility of the self-responsibility of each ward in waste disposal.

At the same time, however, IWWD was normatively undeniable for them as this idea was closely associated with the autonomy of the 23 wards. As noted earlier, autonomy expansion was the primary political goal for the 23 wards. IWWD, which required the self-responsibility of each ward for the management of its own waste, was tightly connected with the value which had underlain this political movement; IWWD was undeniable for those who had longed to become fully-fledged, independent local municipalities. When IWWD was proposed by Koto ward, the restoration of the public election for ward mayor was on the agenda and the autonomy expansion movement was reaching one of the culminations in its history. In this political context, the wards could not reject the idea that each ward should take a certain role to resolve waste problems. To become independent local municipalities, both in name and in reality, the wards were no longer allowed to remain unconcerned with disposal of their own waste.

Furthermore, although the wards were not willing to accept the institutional responsibility in waste disposal, they rather wanted to have incinerators in their own districts to realise the devolution of collection and transportation. The 23 wards had demanded the devolution in the autonomy expansion movement, although it was limited to collection and transportation. The devolution of the collection and transportation had been prevented in part due to the shortage and uneven distribution of incinerators. The union claimed that a regional management was necessary to make frequent adjustments on how much waste should be transported to which incinerator. Given the shortage and uneven distribution of incinerators, the destination of the waste of a ward had to change frequently to respond to the daily fluctuation in the generation of waste, as well as the overhauls and unpredictable accidents of incinerators. Siting more incinerators dispersedly, therefore, was desirable to attain their political goal.

Thus, IWWD was perplexing for the wards. The idea of distributive justice which concerned only the distribution of waste disposal facilities was not convincing for those who had suffered the burden of other types of locally unwanted land uses. The feasibility problem of siting incinerators in every ward likewise undermined the cognitive legitimacy of this idea; the wards where lands for incinerators were supposedly difficult to find did not want to take on that responsibility, preferring the regional disposal to the local disposal. On the other hand, taking certain responsibility in waste disposal was normatively compelling, as it resonated with the greater autonomy of each ward which they had strongly advocated. Having enough incineration capacity distributed evenly among the wards would promote the devolution of waste collection and transportation, which was also a significant part of the autonomy expansion movement.

In the end, the wards supported the incinerator siting in general, but did not want to take the administrative responsibility in the disposal of their own waste. Despite the conflicting views on distributive justice, they needed to show their commitment to incinerator siting due to the autonomy expansion as well as the pressure from Koto ward, while the self-responsibility of each ward in waste disposal was unacceptable due to feasibility problems. As a result, all they could promise was to help the TMG to implement the incinerator projects.

In sum, the union, the TMG, and the wards were all negative towards the idea that each ward should be institutionally responsible for its own waste disposal, while all of them supported facilitating the incinerators construction. Even Koto ward did not persist in devolving the responsibility, as long as the incineration capacity would increase and the amount of waste into the landfills would decrease. This partial advocacy explains why the strength of the idea of the wards' responsibility was limited in the 1970s while OWOI became influential.

Conclusion

IWWD became influential in policies in the early 1970s as all of the four variables – ideational legitimacy, interests, power of carriers, and exogenous environments – worked positively for this idea.

IWWD was perceived as normatively legitimate because this idea of distributive justice resonated with the underlying values of the major actors, i.e. the efficiency in garbage transportation and the autonomy of each ward. IWWD was consistent with these values as all of them would be realised by a large number of small incinerators evenly distributed among the wards. IWWD was accepted by the TMG as the idea of siting incinerators in every ward would enhance these two values which underpinned the old siting scheme, rather than challenge them. The 23 wards could not deny the idea as it resonated with the autonomy of each ward, which they advocated in the autonomy expansion movement.

Siting incinerators in every ward was cognitively appealing as well. IWWD was perceived by the TMG as a policy solution to the garbage crisis. As the amount of waste was growing beyond expectation, the TMG could not afford to exclude the central part of the 23 wards from the siting of incinerators in order to achieve AWI. Although the practicability of IWWD was doubted, the TMG advocated this idea to facilitate the construction of incinerators necessary for this policy goal.

IWWD was backed by the powerful interest of Koto ward. Koto had a strong interest in facilitating incinerator construction because it believed that the shortage of incinerator capacity was the main cause of the disproportionate burden on the ward. Furthermore, Koto ward was able to influence the policy-making process through its threat of a blockade and the de facto veto power over the new landfills, given Tokyo's over-dependence on the coastal landfill. This interest and power of Koto ward enabled IWWD to be quickly adopted in the governmental policies.

OWOI was congruent with interests of other actors as well, or at least did not raise intense opposition to it. The union recognised that investing in waste disposal facilities would help them resolve the discrimination against cleaning workers and improve their working environment. The wards also had an interest in facilitating the incinerators' construction to promote the autonomy expansion. Thus, all of the major parties supported IWWD to facilitate the incinerator construction due to its ideational legitimacy and its congruence with their interests.

This does not mean that IWWD did not raise any opposition however. IWWD met persistent opposition from the neighbours around the incinerator projects in the implementation process. Although the wards agreed on the projects in principle, the neighbours opposed them as they would disturb local environments. However, this difficulty in the implementation had yet to affect the dominance of the idea in the policies during this period. The impact of the feedback from the implementation will be detailed in the next chapter.

The exogenous environments, i.e. the rapid economic growth and incinerationism, were centred on the positive interaction between the variables for the idea of siting incinerators in every ward. It was this rapid growth of the economy that supported the interest and power of Koto ward. The economic boom resulted in a huge increase of garbage production. The burden of waste disposal on Koto ward had been accumulating under this growing amount of waste. At the same time, it was under the deepening landfill crisis caused by the soaring waste production that the garbage blockade and the political veto of the new landfills worked as bargaining chips in its negotiation with the TMG. Furthermore, IWWD was recognised as a solution to the garbage crisis given the urgent necessity for more incinerators, resulting from the interaction between the rapid waste growth and incinerationism as the policy paradigm in waste management.

Thus, all of the four variables worked harmoniously for the idea of siting incinerators in every ward. The exogenous environments were central in the interaction between them, as cognitive legitimacy, interests and power of the claimant were significantly supported by rapid economic growth and incinerationism. Accordingly, changes in these environments significantly affected the strength of IWWD as will be detailed in the following chapters.

On the other hand, the self-responsibility of each ward in waste disposal was not reflected in the relevant policies. This was mainly because this requirement was against the union's interest. The union quickly reacted to the idea and made sure that IWWD would not influence the devolution of waste management to the individual ward. The union was able to make its claim heard by the TMG, because the Minobe administration was a socialist government and hence could not ignore the claim and needed the workers' cooperation to tackle the garbage crisis. Furthermore, the national political environment provided the opportunity for the union to prevent the devolution of waste management.

To make matters worse, devolving waste disposal to each ward was not cognitively convincing to all of the parties, though it was normatively compelling. It was recognised that devolving waste management to each ward would worsen the garbage crisis rather than solve it. The wards did not want the responsibility of incineration to be forced on them, because for some it would be impossible to find an adequate site for an incinerator and therefore incineration was better performed regionally rather than locally.

Taken together, the four variables worked positively for the idea of siting incinerators in every ward, while the self-responsibility of each ward was against the interest of a powerful actor and recognised as problem causing. As a result, IWWD became influential for the siting of incinerators but not for the devolution of waste management.

Notes

1 Jinkai Shori Keikaku [塵芥処理計画] (Toshi Keikaku Tōkyō Chihō Iinkai, 1939a).
2 Shuto Kensetsu Kinkyū Gokanen Keikaku [首都建設緊急5か年計画].
3 Shōkyakujō Kensetsu Jukkanen Keikaku [焼却場建設10か年計画].
4 Tōkyō-to Chōki Keikaku [東京都長期計画] (Tōkyō-to, 1963).
5 As explained later in this chapter, the construction of two large incinerators in the coastal area was planned to achieve All Waste Incineration.

6 The Tamagawa plant in Ota ward and the Itabashi plant were under reconstruction; the Oi plant in Shinagawa ward and the Koto plant were under construction; the Suginami plant was under negotiation with the local opposition.

7 Gomi Taisaku ni kansuru Kihonteki Kangaekata [ゴミ対策に関する基本的考え方] (Tōkyō-to Kikaku Chōsei Shitsu, 1972).

8 Tōkyō-to Chūki Keikaku 1971 [東京都中期計画1971年] (Tōkyō-to Kikaku Chōsei Shitsu, 1972).

9 Chiyoda, Chuo, Minato, Shinjuku, Bunkyo, Taito, Sumida, Meguro, Ota, Shibuya, Nakano, Toshima, and Arakawa.

10 The Kamata plant in Ota ward and the Nippori plant in Arakawa ward were old, small incinerators with only 60 tonnes/day capacity, originally built in 1936 and in 1931 respectively. The Osaki plant in Shinagawa ward, with 30 tonnes/day capacity, was also constructed before World War II. They were all decommissioned in 1973 (Tōkyō-to Seisō Kyoku, 2000).

11 Tōkyō-to Chūki Keikaku 1972 [東京都中期計画1972年] (Tōkyō-to Kikaku Chōsei Shitsu, 1973).

12 Until 1943, Tokyo City was the local government administrating the area of the 23 wards today, while the Tokyo Metropolitan Government (back then, it was not Tokyo-to [東京都], but Tokyo-fu [東京府]) played the role of a prefectural government (regional government) over the area of the 23 wards and Western Tokyo. However, the Tokyo City Government was merged into the Tokyo Metropolitan Government in 1943, and the latter took over the role of the city government (local government) in the area of the 23 wards.

13 Although the 23 wards were granted status of basic local municipalities similar to cities, they had been categorised not into normal municipalities, but into special local municipalities.

14 This was because of the shift in the occupation policy from democratisation to rapid economic recovery in order to fight against the communists under the political context of the Cold War. As public election was abolished, a mayor was selected by a ward council and needed to be approved by the TMG. Furthermore, the 23 wards had to accept staff dispatched from the TMG. They were also deprived of the administration of some public services in this amendment.

15 Obutsu Sōji Hō [汚物掃除法].

16 Back then, waste disposal stagnated under a privatised disposal system in which private dealers picked up only valuable materials, thereby leaving worthless waste and causing sanitary problems such as cholera (Tōkyō-to Seisō Kyoku, 2000).

17 The Cleaning Act (Seisō Hō [清掃法]) in 1954 and the Waste Disposal and Public Cleaning Act (Haikibutsu no Shori oyobi Seisō ni Kansuru Hōritsu [廃棄物の処理及び清掃に関する法律]) in 1970 also designated the liability to a municipal government.

18 The TMG had performed domestic waste disposal since the Tokyo City Government, which had governed the area of the 23 wards, was merged into the TMG in 1943 in the name of efficiency to win World War II. Since then, waste-related services had been performed by the TMG. The exception was at the end of World War II. The services were delegated to each ward in July 1945 because of the decreasing population as people evacuated from Tokyo to avoid air raids. The TMG resumed to administrate the services again in May 1946 by GHQ's request (Tōkyō-to Seisō Kyoku, 2000).

19 For instance, see Tōkyō-to Gyōzaisei Tantō Semmon Iinkai (1970), Dai Jūyoji Chihō Seido Chōsakai (1970), Dai Jūgoji Chihō Seido Chōsakai (1972), and Tōkyō-to Gomi Taisaku Semmon Iinkai (1972b).

20 The report stated that; "The 23 wards of Tokyo were becoming more independent autonomies. The garbage war was caused by the resistance of an independent ward to which the disposal of waste has been imposed from the other wards. Indifference and dependence on other wards in waste disposal was no longer acceptable if the

wards aspired to become independent autonomies in both name and reality. All Waste Incineration should be achieved by In-Ward Waste Disposal to show the willingness of the people in the 23 wards to take care of their own waste" (Tōkyō-to Gomi Taisaku Semmon Iinkai, 1972b).

21 Haikibutsu no Shori oyobi Seisō ni Kansuru Hōritsu [廃棄物の処理及び清掃に関する法律].

22 The landfills were not located in Koto ward, but next to the ward.

23 A large part of Koto ward today was formed by land reclamation.

24 Shuto Ken Seibi Keikaku [首都圏整備計画].

25 Shotoku Baizō Keikaku [所得倍増計画].

26 Jūdai na Ketsui [重大な決意]. It aimed to block garbage trucks driving through the ward (Kōtō-ku-gikai, 1971c).

27 It is true that there were ideas similar to IWWD even before Koto conceived of IWWD. As explained later in this chapter, IWWD was to extend the values which had underpinned the previous siting scheme. In this sense, Koto did not invent this idea of distributive justice from scratch. But it was Koto ward that formulated and concretised the idea. As Carstensen (2015) maintains, ideas do not emerge from an absolute origin but instead are created when a set of ideational elements are yoked together by political actor.

28 As discussed in the following chapters, Koto ward's attitude towards IWWD changed over time. See Nakazawa (2017) for more details on how Koto ward's stance on IWWD changed in conjunction with economic, policy, and political circumstances.

29 Incineration is a technology to burn waste at a high temperature, while landfill is a way of waste disposal by burying waste under the earth. These have been the main waste disposal technologies in Japan. As mentioned in the next section, incineration had been regarded as the best technology in waste disposal and the shift to an incineration-centred disposal system had been the goal of waste management policy in Japan.

30 This figure is the percentage of incineration in the overall waste disposal including landfilling.

31 It is noteworthy that even Koto did not believe that IWWD was totally achievable. As explained in the next chapter, Koto recognised that it was hardly possible to site an incinerator in every ward due to little land availability in the central part of Tokyo (Yomiuri Shimbun, 1971b).

32 Among wards in the north-eastern part of Tokyo including Koto, there were complaints that they had been left out of the rapid development of Tokyo through the Tokyo Olympics which were held in 1964. Five wards (Sumida, Koto, Adachi, Katsushika, and Edogawa) formed the League of Five Wards for Promoting Development (Kōtō Goku Kaihatsu Sokushin Remmei [江東五区開発促進連盟]) in 1965 and demanded the construction of basic infrastructures in the area (Kōtō-ku, 1965b).

33 They included handing over lands owned by the TMG for new parks, using the site of a factory for the redevelopment project of the ward, considering the construction of a hospital and paying for a civic hall on land owned by the TMG (Tōkyō-to, 1971c).

34 Landfills were an ambivalent matter for the development of the ward. As a large portion of the reclaimed land in Tokyo Bay had been incorporated into the ward, landfills were a symbol of development of the ward, extending its territory on which the future of the ward would be built (Kōtō-ku, 1958, 1960). At the same time, the dumping of garbage in the process of reclamation caused environmental degradation in the ward and hindered its development.

35 The blockades were put into action as a response to the heightened anti-dumping movement in Edagawa, a part of Koto ward where garbage trucks passed through. Among the people in Koto, there had been widespread complaints against the disproportionate burden of waste disposal. The rage was about to erupt especially among people in Edagawa. The Koto ward council led the blockades by itself as it was afraid that the situation would get out of control if they had been carried out by the local people directly (Komatsuzaki Gunji Denki Kankō Kai, 1981).

36 The battle over the Suginami plant will be detailed in the next chapter.
37 Kōyū Suimen Umetate Hō [公有水面埋立法].
38 The ownership was to be decided after the reclamation was completed.
39 The ward also demanded the realisation of dispersed dumping and the approval of local people in Toyosu and Shinonome where garbage trucks would drive through if the inner seawall landfill were built.
40 In Adachi and Katsushika, the neighbours protested against the proposed reconstruction and expansion of the existing plants.
41 An example was an advertisement that the TMG put on major newspapers on 15 February 1972. The article, titled "Waste from a community be disposed of in that community", showed how much waste was emitted and disposed of in each ward.
42 In fact, most of the incinerators were sited along the 7th and 8th beltways.
43 A statement of Shiro Honda from Fukagawa ward (Tōkyō-shi-kai, 1939).
44 The plan argued that it would make the transportation efficient, minimise the trouble for the neighbours and fit with the notion of autonomy that one's garbage should be taken care of by itself.
45 The 35 wards were re-organised into the 23 wards in 1947.
46 Waste at the centre had been incinerated at Fukagawa plants. However, the incinerators were sold after the war (Tōkyō-to Seisō Kyoku, 2000). Consequently, the TMG had planned to dispose of waste from the central area mainly by coastal landfills until the 1960s.
47 For instance, see Tōkyō-to-gikai (1959a, 1959b, 1960a, 1960c, 1961, 1962, 1963, 1965).
48 For instance, see Tōkyō-to-gikai (1959d, 1969, 1971a).
49 For instance, see Suginami Seisō Kōjō Kensetsu Suishin Hombu (1974).
50 What waste should be incinerated was dependent on technologies and policies, and changed from one period to another. For instance, food scraps were not to be incinerated until the early 1960s. Old incinerators were not supposed to dispose of food scraps which contained much water (Tōkyō-to Seisō Kyoku, 2000).
51 Seisō Jigyō no Shōrai Keikaku [清掃事業の将来計画] (Tōkyō-to Seisō Kyoku, 1960).
52 This is the percentage of incineration in the overall waste disposal. On the other hand, All Waste Incineration means that all of the waste meant for incineration be incinerated.
53 This was related to the change in waste separation. Incineration of food waste, which had been dumped in the landfills as it contained too much water to burn, started in 1961. The Tokyo Long-term Plan in 1963 revised the Ten Years Incinerators Construction Plan and intended to expand the incineration capacity in response to this change in waste separation (Tōkyō-to Seisō Kyoku, 2000).
54 The first five-year plan mainly concerned human waste disposal. Incineration was focused upon from the second five-year plan (Yagi, 2004).
55 The percentage rose to 70% by 1990 and reached 79% in 2011 (Kankyō Shō, 2013; Kojima, 2003).
56 This was partly because Tokyo could rely on coastal landfills (Tōkyō-to Seisō Kyoku, 2000).
57 Another reason the realisation of AWI was delayed was that the existing incinerators became obsolete faster than expected and needed to be renovated, due to change in the quality of garbage, stricter regulations on urban environment, and rapid technological development in incineration (Tōkyō-to Seisō Kyoku, 1971).
58 In fact, IWWD was always associated with AWI. For example, see Suginami Seisō Kōjō Kensetsu Suishin Hombu (1974) and Tōkyō-to Gomi Taisaku Hombu (1974).
59 For instance, see Tōkyō-to-gikai (1959d, 1969, 1971a).
60 The TMG was well aware of this difficulty and launched a research on the possibility of underground incinerators. However, it was denied due to technological and safety concerns in the report of the research (Tōkyō-to Gomi Taisaku Semmon Iinkai, 1972b).

61 "Chiiki Shori [地域処理]" in Japanese. In a questionnaire survey conducted by the TMG, there was a question on what was the proper size of a community in terms of waste disposal (Tōkyō-to, 1971a).

62 Besides poor work environment and occupational hazard, among the Japanese society, there had been discrimination and a disdain of cleaning workers who dealt with dirty garbage. For instance, some of them could not even tell their family what job they were doing. Sometimes, telling the truth led to breakups with girlfriends or fiancés. (Osumi, 1972; Shibata, 1961; Yorimoto, 1974).

63 Kikaku Chōsei Shitsuchō [企画調整室長].

64 See the replies to questions from Takashi Kosugi, Daikichi Murai, Fukaya Takashi (Tōkyō-to-gikai, 1971c), Daikichi Murai (Tōkyō-to-gikai, 1971d), Kokichi Okada (Tōkyō-to-gikai, 1972b), Mitsuru Tanaka (Tōkyō-to-gikai, 1973a), and Mamoru Tajima (Tōkyō-to-gikai, 1973b).

65 革新自治体.

66 For instance, Kyoto, Osaka, Kanagawa, Yokohama, Kobe, Nagoya and so forth.

67 In the meeting of the Tokyo Waste Problems Special Advisory Committee, Minobe stated that the waste disposal should be devolved but it was nearly impossible due to the union's protest as well as the collection-transportation which barely managed to be operated with 60% reliance on private business (Asahi Shimbun, 1972b).

68 It has been a common view that labour unions in Japan were less powerful due to their decentralised structure based on enterprise unionism and could not exert significant influence on the national politics as described as "corporatism without labour". On the other hand, some have emphasised Japanese unions' positive roles in enterprises as well as in the national politics. For more details, see Kume (1998).

69 This point will be detailed in the next chapter.

References

Asahi Shimbun. (1971). Hirakinaotta Kōtō-ku. *Asahi Shimbun Sha, 26.9.1971*.

Asahi Shimbun. (1972a). Gomi Jikunaishori ni Kyōryoku Yōsei. *Asahi Shimbun Sha, 4.2.1972*.

Asahi Shimbun. (1972b). Ku e Ikan Ima wa Muri. *Asahi Shimbun Sha, 14.10.1972*.

Carstensen, M. B. (2015). Conceptualising Ideational Novelty: A Relational Approach. *The British Journal of Politics & International Relations, 17*(2), 284–297.

Dai Jūgoji Chihō Seido Chōsakai. (1972). *Dai Jūgozi Chihō Seido Chōsakai Tōshin*.

Dai Jūyoji Chihō Seido Chōsakai. (1970). *Dai Jūyoji Chihō Seido Chōsakai Tōshin*.

Ishii, A. (2006). Tōkyō Gomi Sensō wa Naze Okottanoka. *Haikibutsu Gakkai Shi, 17*(6), 340–348.

Ito, Y. (1982). *Edo no Yume no Shima*. Tokyo: Yoshikawa Kōbun Kan.

Kankyō Shō. (2013). *Ippan Haikibutsu Shorijigyō Jittai Chōsa no Kekka*.

Keizai Kikaku Chō. (1956). *Keizai Hakusho: Nihon Keizai no Seichō to Kindaika*.

Kojima, N. (ed.). (2003). *Gomi no Hyakka Jiten*. Tokyo: Maruzen.

Komatsuzaki Gunji Denki Kankō Kai (ed.). (1981). *Shikon: Komatsuzaki Gunji Sono Hito to Gyōseki*. Tokyo.

Kōtō-ku. (1958). Minami ni Nobiru Kōtō-ku. *Kōtō-Ku-Hō Dai 190 Gō*.

Kōtō-ku. (1960). Nobiyuku Kōtō-ku. *Kōtō-Ku-Hō Dai 233 Gō*.

Kōtō-ku. (1965a). *Ijō Hassei: Hae Bokumetsu Taisaku no Kiroku*.

Kōtō-ku. (1965b). Toshin tono Kakusa o Nakusō. *Kōtō-Ku-Hō Dai 330 Gō*.

Kōtō-ku. (1974). *Gomi Mondai to Kōtō-ku*.

Kōtō-ku, & Kōtō-ku-gikai. (1970). Jūyon Gō Umetatechi (Yume no Shima) ni Kakawaru Yōbō Shisetsu ni tsuite.

Kōtō-ku-gikai. (1971a). Gomi Tōki Hantai ni Kansuru Kōkai Shitsumonjō.

Kōtō-ku-gikai. (1971b). Nijūniku ni Taisuru "Gomi Tōki Hantai ni Kansuru Kōkai Shitsumonjō" eno Kaitō Jōkyō.

Kōtō-ku-gikai. (1971c). Seimei.

Kume, I. (1998). *Disparaged Success: Labor Politics in Postwar Japan.* New York: Cornell University Press.

Mainichi Shimbun. (1971a). Kankyō o Mamoru tame Gomisute wa Gomen. *Mainichi Shimbun Sha, 7.9.1971.*

Mainichi Shimbun. (1971b). Niku "Gomottomo", ato Rokku "Mōsukoshi Matte". *Mainichi Shimbun Sha, 6.10.1971.*

Mainichi Shimbun. (1972). Nennai ni Yōchi Kakuho o. *Mainichi Shimbun Sha, 4.2.1972.*

Mainichi Shimbun. (1973). Jikunaishori Gomottomo · · · · . *Mainichi Shimbun Sha, 24.5.1973.*

Nakazawa, T. (2017). A Struggle for Distributive Fairness in Waste Disposal: Koto Ward and In-Ward Waste Disposal in the 23 Wards of Tokyo. *Local Environment,* 22(2), 225–239.

Osumi, S. (1972). *Gomi Sensō.* Tokyo: Gakuyō Shobō.

Shibata, T. (1961). *Nihon no Seisō Mondai.* Tokyo: Tōkyō Daigaku Shuppan Kai.

Suginami Seisō Kōjō Kensetsu Suishin Hombu. (1974). *Gomi Mondai ni Kansuru Chiji Hatsugen Shū.*

Tōkyō Seisō Rōdōkumiai. (1971). *Dai Sankai Chūō Iinkai Gian to Hōkoku.*

Tōkyō Seisō Rōdōkumiai. (1981). *Tōkyō Seisō Rōdōkumiai Sanjūnen-shi.*

Tōkyō Seisō Rōdōkumiai. (1999). *Tōkyō Seisō Rōdōkumiai Gojūnen-shi.*

Tōkyō-shi-kai. (1939). *Tōkyō-shi-kai-shi: Dai Hachikan.*

Tōkyō-to. (1963). *Tōkyō-to Chōki Keikaku.*

Tōkyō-to. (1969). *Tōkyō-to Chūki Keikaku 1968.* Tokyo.

Tōkyō-to. (1971a). *Gomi Mondai ni Kansuru Seron Chōsa Hōkokusho.*

Tōkyō-to. (1971b). Gomi Tōki Hantai ni Kansuru Shitsumon ni taisuru Kaitōsho.

Tōkyō-to. (1971c). Gomi Tōki Hantai ni Kansuru Shitsumon ni Taisuru Kaitōsho no Hosoku ni tsuite.

Tōkyō-to Gomi Taisaku Hombu. (1973). Kaku Chiku no Ugoki. *Gomi Sensō Shūhō No.73.*

Tōkyō-to Gomi Taisaku Hombu. (1974). Gomi Mondai wa Kaiketsu ni Zenshin Shitsutsu aru. *Gomi Sensō Shūhō No.104.*

Tōkyō-to Gomi Taisaku Semmon Iinkai. (1972a). *Kinkyū Sochi ni tsuite Teigen.*

Tōkyō-to Gomi Taisaku Semmon Iinkai. (1972b). *Tōkyō-to no Gomi Taisaku ni tsuite.*

Tōkyō-to Gyōzaisei Tantō Semmon Iinkai. (1970). *Tokubetsu-ku Shichōson Seido oyobi To no Kuiki o Koeru Kōikigyōsei no Arikata.*

Tōkyō-to Kikaku Chōsei Shitsu. (1971). *Tōkyō-to Chūki Keikaku 1970.*

Tōkyō-to Kikaku Chōsei Shitsu. (1972). *Tōkyō-to Chūki Keikaku 1971.*

Tōkyō-to Kikaku Chōsei Shitsu. (1973). *Tōkyō-to Chūki Keikaku 1972.*

Tōkyō-to Seisō Kyoku. (1952). *Jinkai Shori Gokanen Keikaku 1.*

Tōkyō-to Seisō Kyoku. (1960). *Seisō Jigyō no Shōrai Keikaku.*

Tōkyō-to Seisō Kyoku. (1964). Jūgogō Umetate Shobunjō no Kensetsu ni tsuite.

Tōkyō-to Seisō Kyoku. (1971). *Tōkyō no Gomi.*

Tōkyō-to Seisō Kyoku. (2000). *Tōkyō-to Seisō Jigyō Hyakunen-shi.* Tokyo: Tōkyō-to Kankyō Seibi Kōsha.

Tōkyō-to-gikai. (1959a). *Tōkyō-to-gikai Eisei Keizai Seisō Iinkai Sokkiroku: Shōwa Sanjūyonen Dai Ichigō.*

Tōkyō-to-gikai. (1959b). Tōkyō-to-gikai Eisei Keizai Seisō Iinkai Sokkiroku: Shōwa Sanjūyonen Dai Jūgō.

Tōkyō-to-gikai. (1959c). Tōkyō-to-gikai Eisei Keizai Seisō Iinkai Sokkiroku: Shōwa Sanjūyonen Dai Jūgogō.

Tōkyō-to-gikai. (1959d). Tōkyō-to-gikai Eisei Keizai Seisō Iinkai Sokkiroku: Shōwa Sanjūyonen Dai Sangō.

Tōkyō-to-gikai. (1960a). Tōkyō-to-gikai Eisei Keizai Seisō Iinkai Sokkiroku: Shōwa Sanjūgonen Dai Gogō.

Tōkyō-to-gikai. (1960b). Tōkyō-to-gikai Eisei Keizai Seisō Iinkai Sokkiroku: Shōwa Sanjūgonen Dai Ichigō.

Tōkyō-to-gikai. (1960c). Tōkyō-to-gikai Eisei Keizai Seisō Iinkai Sokkiroku: Shōwa Sanjūgonen Dai Rokugō.

Tōkyō-to-gikai. (1961). Tōkyō-to-gikai Eisei Keizai Seisō Iinkai Sokkiroku: Shōwa Sanjūrokunen Dai Jūrokugō.

Tōkyō-to-gikai. (1962). Tōkyō-to-gikai Eisei Keizai Seisō Iinkai Sokkiroku: Shōwa Sanjūnananen Dai Jūgō.

Tōkyō-to-gikai. (1963). Tōkyō-to-gikai Eisei Keizai Seisō Iinkai Sokkiroku: Shōwa Sanjūhachinen Dai Nanagō.

Tōkyō-to-gikai. (1965). Tōkyō-to-gikai Eisei Keizai Seisō Iinkai Sokkiroku: Shōwa Yonjūnen Dai Nijūnigō.

Tōkyō-to-gikai. (1969). Tōkyō-to-gikai Eisei Keizai Seisō Iinkai Sokkiroku: Shōwa Yonjūyonen Dai Nanagō.

Tōkyō-to-gikai. (1971a). Tōkyō-to-gikai Eisei Keizai Seisō Iinkai Sokkiroku: Shōwa Yonjūrokunen Dai Jūgogō.

Tōkyō-to-gikai. (1971b). Tōkyō-to-gikai Eisei Keizai Seisō Iinkai Sokkiroku: Shōwa Yonjūrokunen Dai Sanjūgogō.

Tōkyō-to-gikai. (1971c). Tōkyō-to-gikai Kaigiroku Shōwa Yonjūrokunen Dai Sankai Teireikai.

Tōkyō-to-gikai. (1971d). Tōkyō-to-gikai Kaigiroku Shōwa Yonjūrokunen Dai Yonkai Teireikai.

Tōkyō-to-gikai. (1972a). Tōkyō-to-gikai Eisei Keizai Seisō Iinkai Sokkiroku: Shōwa Yonjūnananen Dai Nigō.

Tōkyō-to-gikai. (1972b). Tōkyō-to-gikai Kaigiroku Shōwa Yonjūnananen Dai Ikkai Teireikai.

Tōkyō-to-gikai. (1973a). Tōkyō-to-gikai Kaigiroku Shōwa Yonjūhachinen Dai Ikkai Teireikai.

Tōkyō-to-gikai. (1973b). Tōkyō-to-gikai Kaigiroku Shōwa Yonjūhachinen Dai Sankai Teireikai.

Toshi Keikaku Tōkyō Chihō Iinkai. (1939a). *Jinkai Shori Keikaku.*

Toshi Keikaku Tōkyō Chihō Iinkai. (1939b). Tōkyō Toshi Keikaku Jinkai Shōkyakujō Keikaku Gaiyō. *Toshi Kōron, 22,* 37–49.

Toshi Keikaku Tōkyō Chihō Iinkai. (1939c). Toshi Keikaku Tōkyō Chihō Iinkai Giji Sokkiroku: Dai Jūsangō.

Yagi, S. (2004). *Haikibutsu no Gyōzaisei Shisutemu.* Tokyo: Yūhikaku.

Yomiuri Shimbun. (1971a). Kankei Kuchō ni Iken Kiku. *Yomiuri Shimbun Sha, 26.11.1971.*

Yomiuri Shimbun. (1971b). Kōtō no Shitsumonjō ni Jūnanaku ga Kaitō. *Yomiuri Shimbun Sha, 12.10.1971.*

Yomiuri Shimbun. (1971c). Kōtō-ku-kai no Kōkai Shitsumon: Kaku Ku no Tenouchi wa. *Yomiuri Shimbun Sha, 6.10.1971.*

Yomiuri Shimbun. (1972a). Kyōryoku wa Suru ga Shōkōjō Yōchi Doko ni Aru. *Yomiuri Shimbun Sha, 13.1.1972.*

Yomiuri Shimbun. (1972b). To no Seisō Mondai-kon de Ku gawa Jikunaishori o Ryōshō. *Yomiuri Shimbun Sha, 4.2.1972.*

Yorimoto, K. (1974). *Gomi Sensō.* Tokyo: Nihon Keizai Shimbun Sha.

4   Declining influence of IWWD

The influence of In-Ward Waste Disposal (IWWD) in siting policies declined through the latter half of the 1970s and the idea was almost forgotten in the 1980s. The completion of One Ward One Incinerator (OWOI) was postponed in 1974 and the principle was compromised in the 1976 revision. By the beginning of the 1980s, the siting of incinerators went back to the old scheme in which facilities were sited in the surrounding area of the 23 wards while the central part was excused. IWWD lost its dominance in the siting policies. On the other hand, the idea of the institutional responsibility of each ward remained weak in influence through this period. Although the devolution of waste management was discussed, the argument was still mostly limited to collection and transportation. Even this partial devolution was not realised. To sum up, the prominence of IWWD as an idea underpinning policy fell from the previous period.

This chapter explores what caused this decline of IWWD. The first section explains the impact of the feedback from the implementation process which led to the 1974 revision, followed by an analysis of the revision in 1976 and the siting policies in the 1980s. Then, the fourth section examines why the idea of the institutional responsibility of each ward was not influential during this period.

Negative feedback from the implementation

Slowdown in the Tokyo Mid-Term Plan 1974

IWWD's strength started weakening in the Tokyo Mid-Term Plan 1974.[1] While the Tokyo Mid-Term Plan 1972 schemed to site 13 new incinerators in the wards without sufficient incineration capacity by 1975, none of the 13 incinerator projects were scheduled to be completed in this three-year plan. As a result, the total incineration capacity in 1976 was planned to be only 11,000 tonnes/day, whereas it was initially projected to reach 17,510 tonnes/day by 1975 in the 1972 plan. Thus, compared to the previous period, the 1974 siting plan slowed down the incinerator siting and started receding from the OWOI policy. Although this plan still held IWWD as the main principle in waste disposal and stated that this idea was taking root among the citizens of Tokyo, IWWD became no more than an abstract goal for the future.

Difficulty in finding sites

The negative feedback from the implementation constrained the influence of IWWD in the Tokyo Mid-Term Plan 1974. As noted in the previous chapter, appropriate sites for the incinerators were difficult to find, particularly in the central part of the 23 wards. Even if a ward was supportive of the project, it was challenging to secure a proper site for it. Thus, limited land availability hindered the progress of the 13 incinerators plan. In June 1972, the TMG explained in the assembly that there were promising candidate sites in only five out of the 13 wards (Tōkyō-to-gikai, 1972b). Half a year after the 13 incinerators plan was announced, no specific candidate site had been publicly announced.

Even at the time when the Tokyo Mid-Term Plan 1974 was issued, only three new sites had been officially publicised. One was the Shinjuku plant which was announced in September 1973 on a piece of land owned by the TMG at Shinjuku Sub-Urban Centre where skyscrapers abounded and more were to be built. Another was in Shibuya ward. Asked by the TMG to cooperate in finding a site, the ward set up the Shibuya Citizen's Committee[2] to select sites for the project. In May 1974, the committee decided on a part of Yoyogi Park as one candidate site. The other was in Meguro, decided in February 1974, at the site of the National Industrial Laboratory which was planned to be relocated.

Nonetheless, it was taking time for the projects in Shibuya and Meguro to be approved due to necessary adjustments with other urban plans. The site in Yoyogi Park needed an alternate park site to make up for the loss of the park space. The project in Meguro ward was also postponed because the site was not available due to the relocation of the laboratory being delayed. As a result, after nearly three years since the declaration of IWWD by Koto ward, only one site was secured for the 13 incinerators project. Limited land availability was a fetter for the progress of IWWD.

Local resistance to the projects

Persistent local opposition was another obstacle for the advancement of the OWOI policy.[3] Even if a ward agreed to a project, the neighbours around the site contested against it. Historically, almost all of the incinerators sited in Tokyo more or less had met local resistance.[4] There are records of local opposition to waste disposal facilities in the early 20th century. The Tokyo City Government started searching for a prospective site for an incinerator from 1903. Nine candidate sites came up in ten years; but none of them were realised, mainly due to opposition from neighbours around the candidate sites (Tōkyō-to Seisō Kyoku, 2000).[5] Siting of waste disposal facilities caused a lot of conflicts after World War II as well. An example is the incinerator project in Yaguchi, Ota ward. This project was originally planned in 1939 and the land for it was already purchased. After the disruption by World War II, the project was written in the Ten Years Incinerators Construction Plan in 1957. This siting project met local opposition and took six years to be brought into operation (Ōta-ku-gikai, 2003). The dispute over the incinerator project in Kita ward is another well-known example. Strong local protest rose right after the project was announced in 1961. The issue was brought to court and the agreement on operation and pollution prevention was made according to the recommendation of reconciliation from the court. It was not until 1969 that the plant started operation (Kita-ku-gikai, 1994).

The 13 incinerators were no exception. The project that attracted the most attention among the 13 incinerators was the one in Shinjuku which was regarded as the symbol of IWWD, for the project was planned in one of the most developed urban areas of Tokyo.[6] Governor Minobe stated that this project was meant to change the perception among the citizens that garbage should be taken out of their communities and disposed of elsewhere, and challenge the view that an incinerator could not be sited in developed urban areas (Tōkyō-to Gomi Taisaku Hombu, 1973). If an incinerator could not be built in one of the prime lands in Tokyo, stated the governor, it would have been impossible anywhere else in Tokyo.[7] Thus, the TMG intended to prove that incinerators could be built even in a developed, congested area in order to facilitate the incinerators siting in other places. However, opposition arose immediately from those who had interests in the redevelopment of this area, such as local shopkeepers, residential associations and big businesses.

Another project which met local opposition was Adachi, where the expansion of the existing incinerator from 600 tonnes/day to 1200 tonnes/day went under negotiation when the garbage war broke out.[8] The incinerator was originally planned in the 1939 siting plan. After the disruption by World War II, the project was written in the Ten Years Incinerators Construction Plan in 1957 and the plant started operation in 1964. The reconstruction and expansion was announced in 1970 to achieve AWI by 1975 as the unexpected change in the quality of garbage damaged the plant.[9] However, this reconstruction plan met resistance from the neighbours. Even after IWWD was declared, the locals continued pleading with the Tokyo Metropolitan Assembly against the expansion and demanded the relocation of the incinerator (Asahi Shimbun, 1971; Tōkyō-to Gomi Taisaku Hombu, 1971; Tōkyō-to-gikai, 1972a).[10]

The most crucial conflict in this period was over the Suginami plant in Takaido, Suginami ward. The Suginami plant was also one of the nine incinerators which were originally planned in 1939. In Takaido, the local opposition movement had prevented the siting ever since the plan was announced in 1966, while the Suginami ward council repeatedly petitioned the TMG to construct an incinerator in the ward in the early 1960s.[11] As explained later in this chapter, this local protest in Suginami was powerful and became a significant bottleneck in the overall progress of the incinerator siting projects.

The local opposition occurred mainly because the incinerator projects were against the neighbours' interests. As IWWD aimed to redistribute to the other wards the waste disposal burdens which had been concentrated in Koto ward, the idea was against the interests of those who were asked to share the burden. Local citizens felt that incinerators would bring environmental degradation to the local community, such as pollution from the plant and garbage trucks as well as traffic jams and accidents. Besides the environmental concerns, an incinerator was regarded as a hindrance to local development, damaging the image of a community and occupying land which could otherwise be used for other purposes beneficial to local development. For example, in Suginami, besides the environmental concerns, the residents opposed the incinerator because the project would occupy the land in front of the rail station which could be used for community development (Naito, 2005; Shōyō Kinen Zaidan, 1983).

The incinerator projects were disliked not just by local citizens in general but also by businesses. For instance, it was local and non-local businesses that intensely opposed the project in Shinjuku. Because the redevelopment project had been planned in the area around the site, the incinerator was opposed not just by local businesses such as shopkeepers, but also by big businesses with interests in the redevelopment project such as mega banks, big real estate businesses, and the hotel industry. They formed the Shinjuku New Urban Centre Development Association,[12] which consisted of 12 big companies including ones from zaibatsu, and industrial and financial business conglomerates, such as Mitsui and Sumitomo (Asahi Shimbun, 1973b, 1973c; Mainichi Shimbun, 1973a, 1973b).

Furthermore, IWWD did not resonate with the local residents, although the TMG tried to justify the projects with this idea of distributive justice. This was in part because, although IWWD justified siting an incinerator in every ward, it did not tell where in a ward an incinerator should be sited or how a site should be selected. For example, the opposition in Takaido criticised the project for the unfairness in the site selection process by pointing out the flaw in the legal procedure of the project siting. It argued that Takaido was chosen, even though it was not the most suitable place in Suginami ward, due to the site selecting process being unduly distorted by political power.[13] In 1939 plan, the site was decided in another place in the ward. As this siting plan was disrupted by the war, the TMG started over the site selection for the project and chose Takaido from the 12 candidates. One of the claims made by the opposition movement was that the decision of the site was legally ineffective, because the site in the 1939 plan was already legally authorised back then and not cancelled yet. Furthermore, the movement suspected that Takaido was selected not because the place was best suited for the incinerator, but because it was politically weak as none of the representatives of the Tokyo Metropolitan Assembly and the ward council lived there (Naito, 2005; Shōyō Kinen Zaidan, 1983).[14]

Similarly, the opposition in Shinjuku also criticised the TMG for the lack of dialogue and consultation with the local people before the plan was officially announced (Kishimoto & Yorimoto, 1976). The project in Shinjuku was announced rather hastily; the TMG had to show Koto ward the progress and its determination for IWWD, as other projects were not expected to be materialised soon (Kodama & Yokoyama, 1974). Also, the neighbours opposing the reconstruction in Adachi demanded to relocate the site elsewhere in the ward while it admitted the legitimacy of IWWD. When the incinerator was built, it was surrounded by rice fields. However, by the time the expansion was planned, the neighbouring area had been urbanised rapidly; there were condominiums, schools and houses densely surrounding the plant. The neighbours argued that there were places in the ward more suitable for the incinerator, while agreeing to the idea of taking care of its own waste (Asahi Shimbun, 1971).[15] Although the idea of siting incinerators in every ward was generally recognised as normatively right, it allowed the locals to make counterarguments.

Thus, the incinerator siting met local resistance; the TMG had to overcome opposition to achieve OWOI. Generally speaking, the TMG was able to force their way through local opposition. Although the institutional siting procedure provided opportunities for citizens to express their opinion in the process of the urban planning decision, it was just a formality and did not offer them enough power to have the government call off the project. Even if landowners disagreed, the TMG could exert eminent domain. It is noteworthy that some opposition movements could win conditions to restrict the amount of waste disposed of by the incinerators to the amount generated in that ward, or simply to exclude waste generated outside of the ward.[16] In Suginami, through the court reconciliation, only 600 tonnes/day were allowed in the incinerator although its capacity was 900 tonnes/day. On top of that, bringing in waste outside the ward was prohibited. In Adachi, the local people won the condition that the plant should dispose of no more than 750 tonnes/day while its capacity was 1,000 tonnes/day (Tōkyō-to Gomi Taisaku Hombu, 1972a; Tōkyō-to Seisō Kyoku, 2000).[17] As the incinerators were planned in the cause of IWWD, the TMG had to accept these conditions. However, local opposition movements could not have the TMG cancel the projects.

It is another story if the government and/or the council of a ward objected to the siting of a project. In the process of siting an incinerator, there was an opportunity for a ward to be consulted before the urban planning decision. Also, the TMG usually sounded out the ward before the official announcement. Although this did not give a ward legal veto, the TMG could not ignore the opinion of an independent local municipality, thereby providing a ward political veto power in siting projects. In fact, incinerator projects were all stopped when the government and/or council of a ward expressed disagreement.[18] The incinerator projects in Nakano and Arakawa in the 1970s were withdrawn when the governments opposed them, as explained later in this chapter.[19] In other words, local opposition movements could not stop the incinerator siting forever unless they had the support of their respective government and/or council of the ward.

Nonetheless, the local opposition movements still managed to impede the progress of the incinerator siting, which led to the revision of the siting plans in 1974. In particular, the local opposition in Takaido, Suginami ward, hindered the advancement of the incinerator siting projects overall.

Resistance in Suginami as the bottleneck of IWWD

The local opposition to the Suginami plant was regarded as the bottleneck of IWWD. The TMG recognised that, if the project in Suginami failed, no project would be successful as the determination of the TMG to site incinerators in every ward would lose its credibility (Shōyō Kinen Zaidan, 1983); the Suginami plant was the cornerstone upon which the success of IWWD depended. The TMG had to resolve the Suginami issue before moving on to the other projects in order to show its determination to achieve IWWD.

The opposition movement in Suginami was persistent because most of the concerned land was owned by the core members of the opposition movement (Tokue Shibata, 2006; Yorimoto, 1977).[20] Furthermore, the landowners were solidly united against the project as a half of them were of the same lineage.[21] Around 70% of the candidate site was owned by the Naito family, who had been influential in this area. This solidarity of the landowners made the movement tenacious. It is true that the TMG could have used eminent domain as the last resort to expropriate the land. In fact, the governor once suggested the use of eminent domain right after the declaration of the garbage war. However, he took it back confronting the criticism and promised not to use it. The TMG was afraid that forcing their way through the opposition in Takaido would raise criticism of authoritarianism and make the forthcoming projects in other wards more difficult (Asahi Shimbun, 1972b; Shōyō Kinen Zaidan, 1983).

Instead, the TMG decided to start over the site selection process in the ward, as the opposition criticised the project for using unfair procedure to select Takaido. In 1972, the TMG established the site selection committee which consisted of the ward government, the ward council, and representatives of local organisations such as shopkeepers' associations, parents and teachers' associations, consumer groups and the like. This promise of not using the compulsory acquisition and the resultant start-over of the site selection process delayed the project further.

The deadlock in Suginami irritated Koto ward. The ward imposed a garbage blockade twice to resolve the situation. The opposition to a temporary relay station in Suginami triggered the first blockade at the end of 1972. While the incinerator siting in Takaido had been stuck in a stalemate, the TMG planned to set a temporary relay station at Wadabori park in the ward to deal with waste increase at the end of year.[22] However, this relay station was opposed by local residents. The TMG cancelled the plan there as the construction was prevented by them (Asahi Shimbun, 1972a). Seeing this turn of events, Koto gave notice of blocking waste from Suginami. The members of the Koto council checked and stopped the garbage trucks from Suginami at the entrance of the 15th landfill. Although the blockade was lifted the next day, piles of garbage were left on the streets of Suginami.

The second blockade occurred in May 1973. In April 1973, Koto demanded that the construction of the new landfill at the inner seawall area be stopped, because the project in Suginami did not make progress and none of the 13 incinerators had come to agreement yet, despite the promise made when Koto ward approved the new landfill siting. Koto accused the TMG of breaking the promise while also criticising Suginami of "local egoism"[23]. The delay of the Suginami plant was regarded as the culprit which delayed the achievement of All Waste Incineration (AWI). Facing pressure from Koto, the site selection committee in Suginami had an emergency meeting. However, members of the local opposition movement in Takaido stormed and disturbed the meeting. That was when Koto decided to put up a blockade again. Waste from Suginami was refused for three days, and the ward was buried in piles of garbage once more.

Governor Minobe visited Koto and asked to lift the blockade, promising to resolve the conflict in Suginami by September 1973. The site selection committee decided on the site and chose Takaido again. Given the result, the TMG resumed talks with the local opposition, but could not reach an agreement even past the date promised with Koto. Koto sent an open letter to the TMG and demanded it keep the promise on the new landfill siting and promote the 13 incinerators construction, the failure of which may result in blockading the transportation of waste once more. Under pressure from Koto, Minobe could no longer continue the dialogue with the local opposition; he made up his mind to resume the process of eminent domain.

The considerable public support also encouraged the TMG to use eminent domain. IWWD had gained significant support from the citizens. Since the declaration of the garbage war, this issue attracted much public attention and was intensively covered by the mass media.[24] The TMG put out an advertisement in the major papers and asked for cooperation for IWWD.[25] According to the opinion poll conducted by the TMG in November 1971, one and half months after the garbage war was declared, 61.4% of the respondents supported IWWD (Tōkyō-to, 1971).[26] This figure rose to 83.1% in December 1973 with only 4.9% opposing the idea (Tōkyō-to, 1973). This widespread advocacy for IWWD made the opposition in Suginami look like "local egoism", or painted them as the "villain" who forced their dirty waste on Koto ward. The opinion poll showed that while Koto's refusal to accept waste from other wards was supported by 89.4%, the support for the opposition in Suginami was only 17.8% (Shakai Chōsa Kenkyū Jo, 1973).[27] The use of eminent domain on Suginami, which the TMG once promised to never use, was also widely supported. The opinion poll conducted at the end of 1973 showed that more than 70% agreed with the use of land expropriation (Tōkyō-to, 1973).[28] This public support encouraged Minobe to decide to resume the process of this authoritarian measure.[29]

Seeing Minobe's determination to use eminent domain, the local opposition finally gave in. The opposition showed its willingness to reconcile and the reconciliation process began in the Tokyo district court in April 1974. The dispute in Suginami finally started working out a resolution.[30] However, it had been already two and half years since IWWD and the 13 incinerators project were announced. Although the TMG, together with the power of Koto ward and the wide public support, were able to force their way through the opposition at last, the deadlock in Suginami had delayed the overall progress of IWWD.

Oil shock and further decline of IWWD in the latter half of the 1970s

The 1976 revision

When the conflict in Suginami was resolved at last, the TMG could have proceeded with the 13 incinerators. However, the influence of IWWD declined further in the latter half of 1970s. The turning point came in 1976 when the Waste Management Advisory Committee Report suggested a further withdrawal from IWWD (Tōkyō-to Seisō Shingikai, 1976). In this report, the scheme to site incinerators in every ward was reconsidered. The report did not deny the need for more incinerator capacity, but relaxed the OWOI policy and rather encouraged joint disposal by several wards. It is true that joint disposal by several wards was not totally excluded as an option even in the previous period. The TMG and Koto ward allowed for joint disposal by two or three wards as explained in the third chapter. However, it had been regarded as an exceptional measure. Although IWWD was still regarded as a significant principle and the report stated that new incinerators should be sited preferentially in wards without one, the OWOI policy was compromised.

Receiving this report, the siting policies receded from the idea of siting incinerators in every ward. The Three Years Plan of Administration and Finance of Tokyo[31] in 1976 budgeted only one project, the Suginami plant. The Low Growth Society and Administration of Tokyo[32] in 1978 showed that, after Suginami, only five or six new incinerators were expected to be necessary by 1985 (Tōkyō-to-gikai, 1978). These siting policies did not show any roadmap to site incinerators in every ward. The influence of IWWD became much weaker in the latter half of the 1970s.

In fact, the incinerators siting during this period did not live up to the ideal of IWWD anymore. Table 4.1 lists new incinerator projects proposed in the 1970s and the 1980s and their results (also see Figure 4.1). As this table shows, none of the 13 incinerators project was realised in the latter half of the 1970s. The two symbolic projects among the 13 incinerators siting plan, i.e. Shinjuku and Shibuya, were cancelled. The project in Shinjuku, which Minobe planned as the symbol of IWWD in 1973, was virtually shelved by the beginning of 1976 (Asahi Shimbun, 1976c, 1976d; Mainichi Shimbun, 1976). The plan was given up and faded out of the scene by the end of the 1970s. The project in Shibuya, which Minobe once praised as a model example of IWWD as the site was selected

Table 4.1 New incinerator projects proposed in the 1970s and the 1980s

Projects	Announcement year	Results
Shinjuku Plant*	Announced in 1973	Abandoned by the end of the 1970s
Shibuya Plant*	Candidate site selected in 1974	Abandoned by the end of the 1970s
Meguro Plant*	Candidate site selected in 1974	Started construction in 1988 and operation in 1991
Nakano Plant*	Proposed in 1976	Withdrawn immediately
Arakawa Plant*	Proposed in 1976	Withdrawn immediately
Hikarigaoka Plant in Nerima	Announced in 1977	Started construction in 1980 and operation in 1983
Ota Plant*	Announced in 1982	Started construction in 1987 and operation in 1990
Ariake Plant in Koto	Announced in 1985	Started construction in 1991 and operation in 1995

* denotes projects in the 13 wards
Source: Tōkyō-to Seisō Kyoku (2000) and news articles

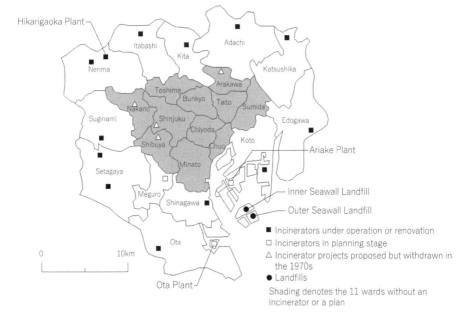

Figure 4.1 Location of incinerators and landfills in 1985.

Adapted from a map of Tokyo by CraftMap (URL: http://www.craftmap.box-i.net/) based on Tōkyō-to Seisō Kyoku (2000).

cooperatively by the ward and the citizens, was cancelled as well. Although the TMG made unofficial overtures to Nakano ward and Arakawa ward in 1976, both of them were withdrawn immediately (Asahi Shimbun, 1976a, 1976b; Mainichi Shimbun, 1977a; Nakano Eki Shūhen Chiku Seibibu, 1991; Tōkyō-to, 1976a). Most of the projects from the 13 incinerators plan failed and were abandoned by the time Minobe stepped down in 1979.[33]

Although the 1976 report recommended that new incinerators necessary for AWI should be built in wards without one, the only project that reached an agreement in the latter half of 1970s was not from the 13 wards, but Hikarigaoka in Nerima ward where one incinerator was already in operation. The Hikarigaoka plant was announced in 1977 and the construction started in 1980.[34] The TMG insisted that this plan was consistent with IWWD as the Shakujii plant, the first incinerator in Nerima, was not sufficient to dispose of all the waste in the ward. However, it was obvious that the TMG was receding from the OWOI policy.

Land availability and local opposition

Land availability and local opposition were partly responsible for the cancellation of some of these projects. The project in Shibuya was cancelled partly because a suitable site was not available. As the Shibuya plant was planned to occupy a part of Yoyogi Park and reduce the park area in the ward, the project needed a replacement site for a park to make up for the loss. However, the prospected alternate site turned out to be unavailable because it was reserved for the New National Theatre (Asahi Shimbun, 1980).[35]

The projects in Shinjuku, Nakano and Arakawa wards were prevented by local opposition. In Shinjuku, local residents and businesses continued to reject the project. In Nakano, the proposal gave rise to a strong objection from the locals, for the proposed site was where a prison was to be removed after a long movement by them for its relocation (Asahi Shimbun, 1976b). The incinerator project was unacceptable to the local movement which had devoted itself to the relocation of the prison for more than 20 years and had already drawn up a blueprint to invite a high school or make a recreational park there. Given this course of events, the mayor and the ward council, who once manifested the support for IWWD, showed a negative reaction to the proposal to site an incinerator there. Arakawa followed a similar course (Asahi Shimbun, 1976a, 1977). In 1976, the TMG proposed an incinerator project, together with a sewage treatment plant, at the site of chemical factories, which were to be relocated due to pollution problems.[36] However, Arakawa ward had already drawn up a blueprint on the development of the site including housing and an evacuation area in case of disaster. The ward council as well as the local residents strongly opposed the proposal. Facing this local opposition, the incinerator plan was withdrawn.[37]

Thus, none of the 13 incinerators materialised in the latter half of the 1970s. The siting of the incinerators conflicted with the local interests in Shinjuku, Nakano and Arakawa. In Shibuya, the limited land availability hindered the project, although the locals were supportive of it. However, local opposition and land availability cannot fully explain why the TMG gave up on these projects without much persistence and started disinvesting from the OWOI policy. The further decline of IWWD in the latter half of the 1970s needs to be understood with reference to fundamental changes in the politics of waste management.

Oil crisis and the consequent financial predicament

The 1973 oil crisis was a main cause of the change in the dominance of the idea in the latter half of the 1970s. The oil crisis, triggered by the 1973 Arab-Israeli War and the oil embargo proclaimed by the Organization of Arab Petroleum Exporting Countries, put tremendous pressure on the Japanese economy and society. The soaring price of oil hit the Japanese economy which had been heavily dependent on the oil imported from the Middle East. With the accelerating rate of inflation, the official interest rate was raised, and capital investment was restrained to control the inflation. Consumer spending dropped rapidly as the Japanese government took policies to restrain the general demand. As a result, 1974 recorded a negative economic growth for the first time since the end of World War II. The Japanese post-war rapid economic growth had come to an end.

The oil crisis and the resultant economic slowdown led to a financial crisis (Jinno, 1995; Mochida, 1995; Tōkyō-to, 1994). The coffers of the TMG had enjoyed a growth in tax revenue due to the long economic boom since 1965.[38] However, as the oil crisis hit the economy, it became rapidly worse off and the budget deficit rose. The TMG's revenue was more dependent on corporation tax than the other municipalities and hence its finance was more sensitive to economic fluctuations.[39] The tax income in 1975 dropped from the previous year for the first time in the post World War II period (Tōkyō-to, 1976b). The actual budget deficit reached 130 billion yen in 1974 and exceeded 200 billion yen in 1977 (Tōkyō-to, 1994); the TMG was about to reach the limit in issuing local government bonds.[40] Governor Minobe declared a "financial war"[41] and pushed forward to reconstructing the finance of the TMG.

This financial crisis reduced TMG's capacity for realising OWOI. Furthermore, incinerators were becoming increasingly expensive due to tightened environmental regulation and technological advances (Tōkyō-to Seisō Kyoku; 1971, 1994, 1995). For instance, the cost of the Suginami plant, which was completed in 1982, was 2.5 times higher than the same-scale plant in Setagaya built in 1969 even after inflation adjustment (Nakasugi, 1982). When its financial capability deteriorated under the economic downturn, the TMG could no longer afford as many incinerators as IWWD required.[42] The financial crisis compelled the TMG to reconsider governmental projects by reassessing their necessity and urgency for them. As the waste growth slowed down under the economic downturn, siting incinerators in every ward was also increasingly seen as excessive.

Widening gap between IWWD and AWI

The oil crisis slowed down the pace of waste increase and undermined the ideational legitimacy of IWWD as a problem-solving idea. Before the first garbage crisis, the amount of waste in Tokyo had increased by around 12% every year on average since 1960. However, in 1973, the amount fell below the previous year for the first time since 1964 as shown in Figure 4.2. Although the amount started increasing again the next year, the pace of the increase slowed down as the economy shifted from rapid growth to low growth; the average growth rate of waste from 1973 to 1978 dropped to around 3.7%.

The TMG started to reconsider the incinerator siting plan when the slowdown of the waste growth was confirmed in 1975. The TMG asked the Waste Management Advisory Committee to review the siting plan with the assumption that the annual increase of waste would drop from 10% to 2–3%. The committee estimated that the amount of waste for incineration would slowly rise to around 12,000 tonnes/day by 1980 and to around 14,000 tonnes/day by 1985 at most,[43] and reported that an additional incineration capacity of 3,500 tonnes/day after the Suginami plant would be necessary to achieve AWI in 1985 (Tōkyō-to Seisō Shingikai, 1976). This figure was far below the estimations made before the oil crisis. In the Tokyo Mid-Term Plan 1970, the amount of waste in 1985 was estimated to be 18,246 tonnes/day (Tōkyō-to Kikaku Chōsei Shitsu, 1971). Based on this estimation, the 13 incinerators plan in the early 1972 aimed at achieving more than 18,200 tonnes/day incineration capacity in total. Even the Tokyo Mid-Term

Figure 4.2   Amount of waste in Tokyo.

Source: Tōkyō Nijūsan-ku Seisō Ichibu Jimu Kumiai (2006, 2013) and Tōkyō-to Seisō Kyoku (2000).

Plan 1974, which modified the calculation formula downward,[44] estimated the amount of waste to be 13,500 tonnes/day in 1980.

The policy rationale for siting incinerators in every ward was undermined by this slowdown of waste increase and the rate of waste generation in the future. As explained in the previous chapter, IWWD was considered to be a policy solution given the enormous pressure to expand incineration capacity to achieve AWI. The end of the rapid economic growth and the resultant slowdown of waste generation alleviated this pressure quickly. Before the oil crisis, IWWD was appealing to the TMG because the amount of waste had been anticipated to increase by 10% every year in the future and incineration capacity could not have kept up with this growth without incinerators being built even in the central area of Tokyo. However, the slowdown of the increase, and more importantly, the estimation that the amount would not grow in the future as quickly as in the past, relieved this pressing need to expand incineration capacity. Rather, siting incinerators in every ward was considered excessive,[45] given the reduced growth of waste.

In other words, IWWD lost its cognitive legitimacy as a problem-solving idea due to the widening gap between IWWD and AWI. As mentioned in chapter three, IWWD was always associated with AWI as the primary policy goal in waste management. The OWOI policy had been supported as the amount of waste was estimated to keep increasing rapidly and more incinerators had been anticipated to be necessary to achieve AWI. However, as AWI was expected to be achieved without as many incinerators as IWWD required, there was no reason for the government to adhere to the idea of siting incinerators in every ward. The TMG stated in the Tokyo Metropolitan Assembly in 1976 that, when the garbage war started, incineration capacity was not expected to keep up with the increase of waste without constructing as many as 13 incinerators by 1980 and that was why it advocated IWWD. However, argued the TMG, constructing 13 incinerators by 1980 became unrealistic given the slowdown of waste increase (Tōkyō-to-gikai, 1976a). For the TMG, IWWD was not a goal but a policy instrument to achieve AWI. Under the slowdown of waste growth, IWWD was increasingly regarded as problematic, rather than as a solution.

Thus, the advent of the low economic growth era after the oil crisis not only weakened the TMG in the implementation of the OWOI policy, but also made IWWD less appealing to the TMG. The TMG could no longer afford as many incinerators as this idea of distributive justice required under the deteriorating financial condition. Moreover, siting incinerators in every ward was regarded as inefficient and excessive, given the financial crisis and the slowdown of waste generation.

Weakening pressure from Koto

On the other hand, the pressure from Koto ward, which had championed IWWD, was eased in this period. Since late 1974, the ward toned down its argument for IWWD. Although the negotiation over the new landfill and other related issues continued, the ward no longer used the term "grave determination", which carried a hint of another blockade of waste in the future. Even the argument for IWWD gradually disappeared in the negotiation with the TMG. In fact, the last time IWWD was referred to in the negotiation was in 1975 in relation to dumping sludge into the landfill in the inner bay area.[46] Even when the incinerator siting was revised in 1976, Koto accepted it without making any persistent complaint.[47] Although the ward bulletin in 1977 proclaimed that the garbage war was not over and the incinerator construction based on IWWD was yet to be achieved (Kōtō-ku, 1977), it was clear that Koto ward did not argue for IWWD as persistently as it had in the previous period.

This was partly because the conflict in Suginami was finally resolved. The conflict in Suginami was the biggest concern not only for the TMG, but also for Koto ward. The mayor of Koto stated that unless Suginami was settled the ward would not stop protesting (Komatsuzaki, Shibata, & Mishiba, 1971). In other words, when the conflict in Suginami started moving toward reconciliation in 1974, Koto had a reason to compromise and end the garbage war which began with Koto's revolt. In fact, right after the anti-incinerator movement in Suginami showed its willingness to reconcile with the TMG in April 1974, the Koto ward council downsized, by more than half, the Garbage Issues Special Committee which had been the headquarter of the garbage war.[48] When the biggest concern of the garbage war was solved, Koto laid down its arms and started finding a way out of the war.

More significantly, the goal that Koto ward pursued by arguing for IWWD was partly fulfilled. Koto argued for IWWD so that the amount of waste into the landfills would be minimised by facilitating the incinerator construction. The incineration rate went up dramatically during 1974–1977 because two new large incinerators, which were planned before the garbage war,[49] and six plants under renovation started operation one after another.[50] The unexpected decrease in waste production due to the oil crisis also contributed to this rise in the incineration rate. The total incineration capacity expanded from 4,340 tonnes/day in 1971 to 6,437 tonnes/day in 1975, and to 11,101 tonnes/day in 1977. Accordingly, the incineration rate also rose from 40.1% in 1972 to 61.8% in 1975. The rate reached 88.2% in 1977 when the reconstruction of the Adachi plant and the Katsushika plant were completed (Tōkyō-to Kikaku Chōsei Shitsu, 1972; Tōkyō-to Seisō Kyoku, 1977; Tōkyō-to Seisō Shingikai, 1976).

In other words, Koto's goal was also to achieve AWI. As noted in the previous chapter, Koto considered the delay of the incinerator construction as the culprit for the environmental degradation of the ward; its protest originated in the campaign against the dumping of un-incinerated food scraps. Koto did not believe that IWWD would be achieved, due to the difficulty in finding appropriate land in the hyper-congested wards (Yomiuri Shimbun, 1971). Nonetheless, Koto made use of the idea to facilitate the incinerator construction and reduce the amount of un-incinerated waste into the landfill. Accordingly, its interest in IWWD became weaker when the incineration rate increased and AWI was expected to be achieved without constructing incinerators in every ward.[51]

Furthermore, Koto ward was losing its power to influence policy making. This was partly due to the disagreement within the ward council. Koto's protest movement was led by the local council. However, the council was not monolithic, even though the members feigned nonpartisan cooperation for this issue. Even right after the garbage war broke out, the council was politically divided into the conservatives, who were keen to use this opportunity to drive the socialist Minobe Administration into a corner, and the progressive parties, who did not feel like hounding the governor too far (Osumi, 1972). When Koto sent the open letter to the TMG in September 1973 and asked for the promises to be fulfilled, the political cracks within the Koto council began to surface, with some insisting on an immediate blockade and others pursuing a political settlement. The council barely came to a decision after more than ten hours of discussion (Mainichi Shimbun, 1973c; Yomiuri Shimbun, 1973).

Besides, as the waste generation slowed down, the garbage crisis dissipated and thereby undercut the negotiating power of the ward. As noted in the previous chapter, Koto ward was powerful as it was able to take advantage of the garbage crisis. When the garbage war broke out, waste management in Tokyo was on the verge of falling apart. The TMG was desperate to secure new landfills to address the rapidly increasing amount of waste. This situation provided the ward with the political opportunity to influence the siting policies by making use of the de facto veto over the new landfill siting. However, since the 1973 oil crisis reduced the waste increase, the sense of urgency in the garbage crisis started fading away.

This led to the settlement of the new landfill siting in the outer bay area, which then further weakened the ward's negotiating power. After giving up the extension of the 15th landfill, the TMG persuaded Koto ward to have the new landfill sited in the inner bay area. However, this landfill was estimated to last only until 1977; the TMG was seeking another landfill in the outer bay area. Although Koto had suspended its approval for the project to pressure the TMG, Koto finally compromised and agreed to the landfill in exchange for two pieces of land owned by the TMG as compensation in 1977 (Mainichi Shimbun, 1977b, 1977c). The new landfill was estimated to last until 1985 at the time, but later the estimated expiry year was extended to 1990 as waste growth further slowed down after the second oil crisis in 1979. Koto lost its bargaining chip to negotiate with the TMG.

Siting policies during the 1980s

Back to the old scheme

IWWD lost its strength further; the idea was no longer referred to in siting policies during the 1980s. The incinerator siting went back to the previous scheme when Minobe left and the new governor, Shunichi Suzuki, came into office in 1979. In the late 1970s, IWWD was still referred to in the siting policies although it was significantly compromised. However, even the words "In-Ward Waste Disposal" disappeared from the policy documents in the 1980s. In the My Town Tokyo[52] in 1981, which was the three-year general plan, IWWD was not even mentioned. While this plan stated that incinerators would be constructed to achieve AWI, only three incinerator projects, Suginami, Hikarigaoka in Nerima, and Meguro, were scheduled in these three years.

Even in the Tokyo Long-Term Plan[53] in 1982, which laid the grand design from 1980 to 1990, only four new incinerators were earmarked for construction, while the reconstruction of three existing ones were planned as their incineration capacity was decreasing due to aging (Tōkyō-to, 1982; Tōkyō-to-gikai, 1982). Consequently, the siting framework shown in this ten-year long-term plan was to leave 11 wards without an incinerator as Figure 4.1 shows. This siting policy was generally succeeded by the Second Long-Term Plan[54] in 1986 which covered the 1986–1995 period. IWWD and OWOI were not even referred to in these plans, let alone the roadmap to site incinerators in every ward. The incinerators siting retrograded to the old scheme in which garbage was to be disposed of at incinerators in the suburbs and the reclaimed lands in the coastal area while the central part was exempted.

Actually, the siting practice in this period did not show any intention to site incinerators in every ward; only three new incinerators were agreed to during the 1980s, besides Suginami and Hikarigaoka in Nerima ward.[55] Two of them were sited in wards in the coastal area where an incinerator had already been in operation: Ota ward and Koto ward.

Ota ward was one of the 13 wards targeted by IWWD, as the existing incinerator was recognised as insufficient to dispose of its own waste. In 1973, the mayor of the ward already showed its willingness to invite an incinerator to the coastal area of the ward for IWWD (Ōta-ku-gikai, 1973). Since then, the ward had urged the TMG to site an incinerator at the reclaimed land in the ward, Keihin Island. Although there were technical conditions to clear, such as the quality of the ground and the height limitation due to the airport nearby, the project was announced in 1982. While some opposition was raised by workers and owners of factories at the island, the incinerator started construction in 1987 (Tsugawa, 1993).

Another was Ariake in Koto ward. This incinerator was planned to dispose of waste in the Coastal Sub-Urban Centre of Tokyo, a newly developing business district at the reclaimed land under the Suzuki administration. Although this incinerator was the second one in Koto ward, no opposition was raised, for the incinerator was meant to self-sufficiently take care of waste generated in this centre; the plant was not accepting waste from outside the area, although this restriction was lifted later. The project was approved in the urban planning decision in 1990 and the construction started in 1991.

The only incinerator sited in a ward without one was Meguro plant, for which the site was already selected in 1974, but the project had been postponed as the negotiation with the central government to acquire the site of the national laboratory was being prolonged (Meguro Seisō Kōjō no Kensetsu ni Hantai Suru Jimoto Yūshi no Kai, 1993). The plant was constructed in 1991 after intense local opposition as explained later in this section.

Thus, IWWD was no longer reflected in the siting policies. Although new incinerators were planned in two of the 13 wards, the other 11 wards were left without any incinerators or even attempts to build one in this period. It was not until the 1990s that IWWD regained prominence in siting policies.

The 1979 oil crisis

Behind this further decline of IWWD was the 1979 oil crisis and the further slowdown of waste production. The 1979 oil crisis, which was triggered by the Iranian Revolution, damaged the Japanese economy which was recovering from the impact of the first oil crisis. Although the impact of the second oil crisis on the Japanese economy and society was moderate compared with the first one (Komine, 2011), the average annual economic growth rate from 1980–1984 dropped to around 3% from around 4.5% during the 1975–1979 period.

The second oil crisis and the resultant economic change restrained the growth of waste in Tokyo. In 1979, the amount of waste generated was reduced by nearly 10% from the previous year; the annual generation of waste, which was 4.4 million tonnes in 1978, dropped to less than 4 million tonnes in 1979 (Figure 4.2). During the early half of the 1980s, the amount decreased gradually to 3.8 million in 1984. This downward trend was reflected in the estimation of future garbage increase in the plans made in the early 1980s. For instance, the Long-Term Plan in 1982 estimated that the amount of waste would only slightly grow to 4.13 million tonnes in 2000 from 4.02 million tonnes in 1980. The necessity to site an incinerator in every ward declined along with this slowing growth in waste production. Given this estimation, the TMG assumed that AWI would be achieved by 1995 with the construction of only five more incinerators including Suginami and Hikarigaoka.[56] The OWOI policy was thought to be even more excessive considering the incineration capacity which the TMG calculated was necessary for AWI.

On the other hand, the financial condition of the TMG was recovering from the crisis in the 1980s. Governor Suzuki who came into office in 1979 immediately readjusted the finance by conducting a thorough restructuring of the expenditure and revenue (Jinno, 1995; Tōkyō-to, 1994). The budget deficit, which had been expanding in the second half of the 1970s, was eliminated by as early as 1981. Increasing tax revenues due to the stable economic growth in the early 1980s helped the recovery, although growth was much slower than the rapid economic growth up to 1973. However, this recovery did not give rise to IWWD because the financial improvement was achieved by cutting investment on infrastructure projects as well as personnel and wages. Governmental projects were strictly assessed by the fiscal efficiency and the necessity of the projects. As the amount of waste stayed constant and AWI was expected to be achieved with the several new incinerator projects, there was no room for the idea of siting incinerators in every ward to regain prominence in the siting policies.

Siting conflict in Meguro

It is worth mentioning here the conflict in Meguro ward, as it illustrates the declining cognitive legitimacy of IWWD. In the early half of the 1970s, IWWD was still recognised as legitimate in general. Although its normative argument was compromised due to its lack of concern with procedures to select a site in a ward, the idea that incinerators should be sited in every ward was generally supported as the need for more incinerators was acknowledged. However, in the conflict in Meguro, the legitimacy of the idea was rejected by the opposition movement.

The opposition denounced the appropriateness of IWWD in two ways. One was that siting incinerators in every ward was excessive given the current and the estimated future amount of waste. The opposition movement argued that OWOI was no longer justifiable because AWI would be achieved without siting incinerators in every ward, as the amount of waste was not increasing then or in the estimated future (Meguro Seisō Kōjō no Kensetsu ni Hantai Suru Jimoto Yūshi no Kai, 1993). While the TMG still tried to justify the project with IWWD, the opposition pointed out that OWOI was no longer reasonable in this situation, questioning the very necessity of a new incinerator in its ward.[57]

The other was a denial of the idea that constructing more incinerators could solve garbage problems. The opposition argued against the notion that more incinerators had to be built as waste was increasing. Instead, it emphasised the importance of the reduction and recycling of waste (Meguro Seisō Kōjō no Kensetsu ni Hantai Suru Jimoto Yūshi no Kai, 1993). In other words, it challenged the policy paradigm which had been dominant in waste disposal, i.e. incinerationism. As IWWD was associated with this policy paradigm, it was challenged by this anti-incinerationism. As detailed in the next chapter, anti-incinerationism spread among local opposition in the 1990s.

Bubble economy

It is also noteworthy here that the amount of waste started increasing again in the latter half of the 1980s. Although the amount of waste in Tokyo slightly decreased after the second oil crisis in 1979, the bubble economy in the late 1980s pushed it up again. The Bank of Japan reduced the official bank rate in response to the sudden appreciation of the Japanese Yen after the Plaza Accord in 1985. The expansionary monetary policy led to the influx of huge speculative money into Japanese stocks, bonds and the real estate market. The Nikkei Stock Index hit its highest point in history in 1989. The average annual growth rate in Tokyo recorded 6.9% in the latter half of the 1980s.[58]

The amount of waste produced per year started increasing again along with this economic boom. In Tokyo, the amount, which was constant at around 3.8 million tonnes during the first half of the 1980s, had grown by 5.6% annually on average since 1985, and reached 4.9 million tonnes in 1989 (Figure 4.2). As the garbage production in Tokyo was sensitive to economic fluctuations, the booming economic activities were accelerating waste generation, especially paper waste due to the rapid spread of office automation and the plummeting price of used paper. From 1984 to 1989, the amount of garbage in Tokyo increased by nearly 30%, while that of the national average was around 16% (Haikibutsu Gakkai, 2003; Tōkyō-to Seisō Kyoku, 2000).

However, this sudden increase of waste had yet to bring back the influence of IWWD on the siting policies, although the increase of waste along with the booming economy was acknowledged. The Second Long-Term Plan issued in 1986 referred to this trend of garbage increase; but the waste growth was underestimated in this policy, forecasting that the amount would stay flat until 2000. Even My Town Tokyo in 1989 did not mention IWWD, although the rapid growth of garbage was recognised as a serious problem. The revival of the idea of distributive justice had to wait until the protest from Koto ward in the 1990s, which was triggered by the new landfill project.

Weak influence on the devolution of waste management

Arguments on the devolution

On the other hand, the influence of the idea of the institutional responsibility of each ward in waste disposal stayed weak in this period. As explained in the previous chapter, IWWD raised the issue on the institutional responsibility of each ward in waste disposal. However, this idea was downplayed through the latter half of the 1970s and the 1980s.

The devolution of waste management was brought up in some governmental reports which concerned the financial reconstruction and/or the autonomy system reform. For instance, the devolution of waste management was mentioned in the report by the Tokyo Administration and Finance Urgent Project Team[59] in 1975. The Tokyo Financial Reform Advisory Committee[60] also proposed devolution. In 1984, the Tokyo Administrative System Investigative Committee,[61] as an advisory committee of the TMG, issued the report on institutional reform of Tokyo, in which waste management was referred to as one of the administrative services that should be devolved from the TMG to each ward.

The devolution of waste management was discussed by the 23 wards as well. The 23 wards were dissatisfied with the autonomy reform in 1974 which restored the public election of the ward mayor but did not grant them the status of a normal local municipality. They consequently set up the Special Wards Administration Investigation Committee[62] to look into further autonomy expansion. The committee had been investigating the Special Wards System since 1974 and its fifth report in 1981 recommended that waste management should be devolved to each ward (Tokubetsu-ku-sei Chōsakai, 1981). In 1986, the TMG and the wards came to an agreement on the "Basic Policy of Tokyo-Wards System Reform",[63] which mentioned waste management as one of the services which should be handed over to the wards.[64]

However, most of these reports concerned only the devolution of collection and transportation; the institutional responsibility of each ward for waste disposal was not considered. The only exception was the Tokyo Administration and Finance Reform Committee, which proposed the devolution of not only collection and transportation but also waste disposal – although not landfilling. Yet, this report also did not call for the self-sufficient disposal of waste by each ward, but regional disposal by a local government association formed by the 23 wards[65] or joint disposal by several wards. The proposal of the report was that the 23 wards should have the legal responsibility for waste disposal, while a regional organisation should perform these services regionally.

Thus, while the devolution of waste management was discussed in the political context of the autonomy expansion movement of the 23 wards, the argument was almost limited to collection and transportation; the influence of the idea of the self-responsibility of each ward in waste disposal was still weak in this period.

Weak cognitive legitimacy and the conflict with the union's interests

The idea that each ward should be institutionally more responsible for its own waste was prevented from influencing the siting policies mainly because the idea was not regarded as cognitively legitimate. As explained in the previous chapter, devolving the responsibility in waste disposal was recognised as unrealistic and all parties believed that waste disposal would be better performed regionally by the TMG, rather than locally by each ward, given the uneven distribution of disposal facilities. This weak cognitive legitimacy limited the argument on the devolution to collection and transportation; no party seriously advocated the self-responsibility of each ward for waste disposal.

In addition, even the devolution of collection and transportation was hard to realise despite the recommendations of these reports, for it was against the organisational interests of the Tokyo Cleaning Workers' Union. The union opposed any devolution of waste management as it would divide and weaken its organisation. The union denounced the devolution as merely a means to cut down the cost of waste management facing the financial crisis and contended that it would not expand the autonomy of the wards.

Every time the governmental consultative committees recommended the devolution, the union immediately took action to prevent it. When it was discovered that the devolution was discussed in the Tokyo Administration and Finance Urgent Project Team in 1975, the union talked to the Bureau of Waste Management of Tokyo and made it admit that waste management was difficult to be performed individually by a ward due to the insufficient waste facilities, the difficulty to develop appropriate waste disposal technologies by a ward, and the uneven distribution of the cleaning staff (Tōkyō Seisō Rōdōkumiai, 1981). When the devolution was discussed in the latter half of 1970s, the union visited the mayors, the political parties, and the unions of the wards in 1977 to make a petition against the institutional reform (Tōkyō Seisō Rōdōkumiai, 1981). The union continued opposing the devolution throughout the 1980s. It organised the One Million Leaflets distribution campaign in 1979 as a response to the intermediate report of the Tokyo Finance Reconstruction Committee which recommended devolving waste management immediately. In 1985, the One Million Signature campaign was launched in response to the "Basic Policy of Tokyo-Wards System Reform". As the Local Autonomy System Investigation Committee[66] was discussing the autonomy reform, given the agreement between the TMG and the 23 wards, the union escalated its activities against the devolution. It appealed to representatives of the National Diet and the local assemblies, the mayors of the wards, residential associations and so forth, by informing them of the problems with the devolution.

Furthermore, the wards did not unanimously support the partial devolution; some of them were reluctant, or rather negative, to accepting even the responsibility in collection and transportation. When the union visited each of the 23 wards in 1977, some of them (i.e. Chuo, Bunkyo, Meguro, Ota, Suginami, Kita, Nerima, and Katsushika) showed lukewarm, or rather negative, attitudes to the devolution while others firmly advocated it (Tōkyō Seisō Rōdōkumiai, 1981). They thought that the devolution was premature, given the insufficient and uneven distribution of facilities necessary for waste management to be performed by each ward. Furthermore, as the TMG advocated the devolution in part for the financial reconstruction, there was a suspicion among the wards that waste management would be devolved without handing over enough revenue to the wards.

Conclusion

The decline of IWWD in 1974 was due to negative feedback concerning its implementation. The difficulty in putting the idea into practice adversely affected its influence on the siting policy. Siting incinerators in every ward was difficult to implement because the TMG lacked sufficient power due to the limited land availability and strong local opposition. The incinerator projects were against the interests of the locals while IWWD was normatively not sufficient to persuade them otherwise. Although the TMG, assisted by the power of Koto ward and wide public support, was able to overcome the local opposition in the end, the progress of the siting projects did not proceed as planned. Confronting the difficulty in the implementation, the TMG had to revise the siting policy.

To make matters worse, the power of the TMG and ideational legitimacy of IWWD were weakened due to the low economic growth caused by the oil crises. When the conflict in Suginami ward was finally resolved, the TMG could have advanced the OWOI policy. However, by that time, the TMG's motive and power for implementing IWWD had been compromised. The financial conditions deteriorated amidst the economic downturn; the TMG could no longer afford as many incinerator projects as IWWD required. IWWD's ideational legitimacy also declined in comparison to the previous period. IWWD was perceived as a problem-solving idea only when it was assumed that the amount of waste would keep increasing rapidly and AWI could not be achieved unless incinerators were constructed in every ward. Accordingly, as the rate of waste generation slowed down due to the low economic growth, siting incinerators in every ward was increasingly recognised as excessive when compared to the capacity necessary for AWI.

The interest and the power of the Koto ward, the original claimant, also became weaker during this period. Koto's interest was in minimising the amount of waste coming into the landfills by facilitating incinerator construction. When the waste growth slowed down, and some incinerators planned before the first garbage war started operation, the goal that Koto ward pursued by claiming IWWD was largely fulfilled. Koto ward was losing its motivation to adhere to IWWD. Furthermore, the diverging interests in the ward council as well as the fading garbage crisis disempowered this claimant. As a result, the pressure exerted by Koto ward to push IWWD alleviated quickly.

Thus, compared with the previous period, the ideational legitimacy as well as the interest and power of carriers declined and worked negatively for IWWD. Exogenous environments played a critical role in the negative interaction between the explanatory variables. The economic downturns triggered by the oil crisis damaged the financial capacity of the TMG to achieve IWWD, undermined the ideational legitimacy and weakened the interest and power of Koto ward. In other words, IWWD was a product of the rapid economic growth. As explained in chapter three, the soaring amount of garbage amidst the rapid economic growth and the resultant landfill crisis gave Koto ward both the motivation to advocate IWWD and the power to influence policies. IWWD's ideational legitimacy and financial feasibility were also dependent on the assumption that the economy would keep increasing rapidly along with the amount of waste and government revenue. Accordingly, IWWD's prominence fell when the era of rapid economic growth came to an end.

Yet, the bubble economy and the resultant garbage growth did not bring back the dominance of IWWD in the latter half of the 1980s. This shows that there was a time gap between the change in the economic condition and its impact on the prominence of IWWD. The response to the exogenous change was rather slow and took time to be reflected in the relevant policies. In addition, the change in external environments is not sufficient for an idea of distributive justice to be influential on the relevant policies; a powerful claimant is necessary. In fact, it was not until Koto ward resumed its campaign in the 1990s that the influence of IWWD was revived.

On the other hand, the idea of the self-responsibility of each ward in waste disposal was not influential on the movement for the devolution. Although the devolution of waste management was discussed, the argument was mostly limited to collection and transportation; no major parties seriously considered the idea that a ward should take the institutional responsibility for its own waste disposal. It was the shared recognition by all the actors that waste disposal needed to be regionally managed and that the local disposal by each ward was not practically feasible; the cognitive legitimacy of the idea was weak. Furthermore, even a partial devolution was prevented by the union whose interest was against this institutional change.

Nonetheless, the self-responsibility of each ward in waste disposal appeared in the policy agenda and became influential in the 1990s. The next chapter explains why this was the case despite this lack of concern in the 1980s.

Notes

1　Tōkyō-to Chūki Keikaku 1974 [東京都中期計画1974年].
2　Shibuya-ku Gomi Shori Taisaku Kumin no Kai [渋谷区ゴミ処理対策区民の会]. This committee consisted of the ward government, the ward council and citizens' groups (Shibuya-ku-gikai, 1976).
3　For more discussion on the responses of ward governments and local residents to IWWD and how these responses changed over time, see Nakazawa (2017).
4　Not only incinerators, but also relay stations were averted by neighbours. There were many relay stations where garbage was transhipped from a small truck to big one, or to a ship. Relay stations had been nuisances for neighbours due to traffic problems, smells, spillover, and so forth, and thereby caused local protests (Tōkyō-to Seisō Kyoku, 2000). In the garbage war, the TMG planned to construct relay stations to reduce the number of garbage trucks driving through Koto ward. However, they met opposition in Shinjuku, Shibuya, Ota, and Adachi.
5　Another example was a conflict between Meguro and Shibuya, which were not a part of Tokyo City then but merged with the city later. The Shibuya town constructed an incinerator in Meguro town. In 1926, the neighbours protested against pollution from the plant. Shibuya and Meguro tried to settle the conflict with compensations and a condition that waste from Meguro would be disposed of by the incinerator as well. However, the people in both of the towns were dissatisfied with the settlement and the incinerator was abandoned in 1931 (Tōkyō-to Seisō Kyoku, 2000).
6　Opposition to the 13 incinerator projects occurred in other places as well. For example, residents in Shibaura, Minato ward, petitioned the Tokyo Metropolitan Assembly against the project (Tōkyō-to-gikai, 1972b). In Shibuya, there were protests from neighbours in the process of selecting a candidate site (Shibuya-ku-gikai, 1976). The project in Meguro went through intense local opposition in the 1980s, as explained later in this chapter (Meguro Seisō Kōjō no Kensetsu ni Hantai Suru Jimoto Yūshi no Kai, 1993).

7 This was called the "Tokonoma theory" [床の間理論]. Tokonoma refers to a built-in recessed space in a traditional Japanese style reception room, where artistic ornaments, such as a hanging scroll, pottery or flower arrangements, are displayed. Minobe used the word as a metaphor for prime land in Tokyo.

8 In the same period, there was a conflict in Katsushika ward over the renovation and expansion of the existing incinerator.

9 Back then, the increase of petrochemical products raised the burning temperature in incinerators and damaged their capacity (Tōkyō-to Seisō Kyoku, 1971).

10 After the intense protest, the reconstruction was brought to an agreement in November 1972 with several conditions: limiting the amount of disposed waste to less than 750 tonnes/day; building routes for garbage trucks; preventing traffic accidents; setting a public space and wooded area surrounding the plant; building welfare facilities; setting a committee to monitor and manage pollution prevention; and so forth (Tōkyō-to Gomi Taisaku Hombu, 1972a). The reconstruction started in 1974 and was completed in 1977.

11 This invitation of an incinerator was initiated by the Suginami Cleaning Association, a local organisation which was founded for the purpose of making a clean and sanitary local environment in response to the campaign against flies and mosquitoes during the 1950s (Shōyō Kinen Zaidan, 1983).

12 Shinjuku Shin Toshin Kaihatsu Kyōgikai [新宿新都心開発協議会].

13 The opposition suspected that the TMG intended to delegitimise the local movement by claiming IWWD (Naito, 2005; Shōyō Kinen Zaidan, 1983; Tōkyō-to Seisō Kyoku, 2000). The leader of the opposition viewed IWWD negatively as a mere slogan to disrupt the opposition in Takaido and argued that political and economic activities such as waste disposal could not be restricted to one ward (Yorimoto, 1977).

14 The suspicion was raised in the court that another place in the ward, Izumi-cho, was chosen before Takaido, but was withdrawn because a high official of the TMG and influential members of the ward council lived there. There was even a rumour that the wife of the high official said that an incinerator would never be allowed there (Naito, 2005; Shōyō Kinen Zaidan, 1983).

15 In the protest against the expansion, they proposed an alternative site in the ward, Toneri (Mainichi Shimbun, 1971b). However, this proposal was rejected by the council of the ward and the TMG (Mainichi Shimbun, 1971a, 1972b).

16 The TMG did not want these restrictions on the operation of the incinerators (interview with an official of Clean Association of Tokyo 23 Ward in 31 October 2011). The TMG adopted IWWD because it thought that this idea would help make garbage transportation more efficient and contribute to achieving AWI. However, garbage transportation might be inefficient if shipment of waste across the wards was prohibited because the nearest incinerator for a certain area in a ward was not necessarily the one in the ward, given the intricate borderlines of the wards. Furthermore, the TMG wanted to build the largest incinerators possible in given lands and incinerate as much waste as possible regardless of where it was generated, in order to achieve AWI as soon as possible and to save landfills. For instance, see Meguro Seisō Kōjō no Kensetsu ni Hantai Suru Jimoto Yūshi no Kai (1993) and Tōkyō-to Seisō Kyoku (1992).

17 Later, the neighbours in Meguro also won a similar agreement, which limited the amount of waste disposed of in the incinerator to the amount generated within the ward (Meguro Seisō Kōjō no Kensetsu ni Hantai Suru Jimoto Yūshi no Kai, 1993; Tōkyō-to Seisō Kyoku, 1992). In Nerima, the Hikarigaoka plant was allowed to operate only one of the two furnaces with 150 tonnes/day capacity each. The operation of the two furnaces was allowed only 80 days per year at most. However, some of these restrictions were modified or lifted in the early 1990s by request from the TMG to deal with the rapid increase of waste during the latter half of 1980s (Tōkyō-to Seisō Kyoku, 2000).

18 The anti-garbage pollution movement in Koto was powerful as it was led by the government and the council of the ward.

19 In the 1990s, the incinerator projects in Meguro, Setagaya, and Koto were cancelled as the wards disagreed with them.
20 It was owned by 16 individuals and 12 of them were core members of the movement (Yorimoto, 1977).
21 For more details of the opposition movement, see Yorimoto (1977).
22 In Japan, it is an annual activity to do a deep house cleaning at the end of the year. Back then, temporary relay stations were built in eight places in the 23 wards to deal with the increased amount of waste at the end of the year. The station in Watabori, Suginami ward, was one of them. Facing the protest there, the station was sited in another place in the area (Tōkyō-to Gomi Taisaku Hombu, 1972b, 1973).
23 "Chiiki Ego [地域エゴ]" in Japanese. It is used pejoratively to refer to local opposition against a facility recognised as necessary for everyone.
24 Teruyoshi Shibata (2001a, 2001b) analyses the role of media in the garbage war and points out the bias of major newspapers favouring IWWD and its influence on public opinion.
25 On 15 February 1972. Entitled "Waste from a community be disposed of in that community", this advertisement highlighted how much waste was generated and disposed of in each ward to illuminate the inequality among the wards.
26 Of the respondents, 40.4% showed strong support while 21.0% showed weak support.
27 Even within Suginami, local groups supporting the incinerator, such as the labour union, the cleaning association, the consumer group, the women's association, the residential association, the association of shopkeepers, and the parents' and teachers' association, gathered and launched an organisation to promote the siting (Asahi Shimbun, 1973a).
28 IWWD accounted for 42.7% of the reasons for supporting the eminent domain, and local egoism for 15.9%.
29 Minobe stated in the new year address in the office that the necessity of incinerators, the resumption of the process of the eminent domain and IWWD were widely supported, referring to the result of this survey (Tōkyō-to Gomi Taisaku Hombu, 1974).
30 It was in 1978 that the construction started after more than ten years of dispute.
31 Tōkyō-to no Gyōzaisei Sankanen Keikaku [東京都の行財政三ヵ年計画] (Tōkyō-to, 1976b).
32 Teiseichō Shakai to Tosei [低成長社会と都政] (Tōkyō-to Seisaku Shitsu, 1978).
33 As noted in the next section, only two of the 13 projects, Meguro and Ota, were realised.
34 The project in Hikarigaoka did not go without a local protest either. Hikarigaoka plant was proposed in the site of Grant Heights, houses for the U.S. Army and its families. During World War II, the lands were forcefully taken from farmers to construct an airport. When the war ended, the U.S. Army came in. In the early 1960s, the neighbours developed a movement and demanded the restoration of the land. The land was returned in 1972 and it was planned to construct there a new town of 12,000 houses with 42,000 residents. The incinerator was proposed to deal with waste in this new town. However, the neighbours around the new town complained that the proposed site was too close to the existing houses, being located at the edge of the new town. They did not oppose the necessity of the incinerator; they called for siting the plant in the centre of the new town. Seeing the local protest, the ward council also sent a request to change the location of the incinerator. As a result, the TMG decided to move the site towards the centre (Nerima-ku-gikai, 1991).
35 This issue on the cancellation of the site in Yoyogi Park was raised in the 1990s again.
36 They are the factories of Asahi Denka Kogyo Kabushiki Gaisha (now known as ADEKA). In the factories, inorganic mercury was used to produce caustic soda. Having gone through the disaster of Minamata Disease, the poisoning by organic mercury in Kumamoto and Niigata prefectures, the Japanese government inspected all factories that used mercury. By this inspection, it was found that Asahi Denka had discharged wastewater containing mercury into the Sumida River. As a result, restric-

tions were imposed on the company on the usage of underground water and expansion of facilities, so it decided to relocate the factories (Arakawa-ku, 1989).

37 While the incinerator was deleted from the picture, the sewage plant was built there.

38 This economic boom was called "Izanagi Keiki".

39 Corporate inhabitant tax and corporate enterprise tax accounted for around 50% of the tax revenue of the TMG (Tōkyō-to, 1994).

40 If the rate of fiscal deficit to a standard fiscal scale (Hyōjun Zaisei Kibo [標準財政規模]) was over 5%, a prefectural government was imposed restriction on issuing bonds unless they handed in a financial reconstruction plan to the Ministry of Home Affairs (for city, town and village governments, the rate was over 20%) (Tōkyō-to Kikaku Chōsei Shitsu, 1975).

41 The war was defined as the battle against the centralised budget structure through which local governments were controlled by the central government (Tōkyō-to, 1976b).

42 The financial reconstruction also required wage cuts and undercut the very political foundation of the Minobe administration which had been supported by the labour unions. The Minobe administration became a lame duck in the second half of the 1970s.

43 The report made the upper and lower estimations.

44 The TMG stated that the amount of waste in the previous plan was overrated because the calculation was made based on the assumption that every garbage truck was loaded to 80% of its capacity, although it was in fact only 60–70% (Asahi Shimbun, 1974).

45 After the oil crisis, IWWD was criticised in the Tokyo Metropolitan Assembly. For instance, Yuji Otsuka repeatedly argued that the incinerator siting plan based on IWWD was excessive and wasteful, given the declining amount of waste (Tōkyō-to-gikai, 1973, 1974a, 1974b, 1975a, 1975b, 1976b).

46 In "Seisō Kōwan Mondai to Kōtō Ku", which collected documents on waste and harbour issues in the ward, the last document in the 1970s which (indirectly) referred to IWWD in the negotiation with the TMG was the "Gesui Shorijō Odei Shōkyaku Zampai Chūō Bōhatei Uchigawa Shobunjō Tōki Mondai ni tsuite." (Kōtō-ku, 1975). In this document, Koto ward required the implementation of the conditions which were agreed upon when the ward approved the landfill.

47 The mayor of Koto was a member of this advisory committee.

48 The number of the committee members was reduced from 23 to 10 (Mainichi Shimbun, 1974).

49 The Oi plant in Shinagawa ward and the Koto plant were newly built and started operation in 1973 and in 1974 respectively.

50 The Tamagawa plant in Ota ward in 1973; the Itabashi plant in 1974; the Katsushika plant in 1976; and the Adachi plant in 1977.

51 The mayor of the ward stated in 1983 that garbage pollution was alleviated compared to the time of the first garbage war as the incinerator siting made progress and AWI was almost achieved (Kōtō-ku-gikai, 1983).

52 Mai Taun Tōkyō 81 [マイタウン東京 81] (Tōkyō-to Kikaku Hōdō Shitsu, 1981).

53 Tōkyō-to Chōki Keikaku: Mai Taun Tōkyō 21 Seiki o Mezashite [東京都長期計画: マイタウン東京 21世紀をめざして] (Tōkyō-to, 1982).

54 Dai Niji Tōkyō-to Chōki Keikaku [第二次東京都長期計画] (Tōkyō-to Kikaku Shingi Shitsu, 1986).

55 The Suginami plant and the Hikarigaoka plant (Nerima) were under construction. They started operation in 1982 and in 1983 respectively.

56 Suginami, Hikarigaoka in Nerima, Meguro, Ota and one undecided (Tōkyō-to, 1982; Tōkyō-to-gikai, 1982).

57 The opposition also argued that the TMG padded the amount of waste by using three types of calculation formulae. "Seisō-tonne [清掃トン]", literally meaning "cleaning-tonne", was the way to estimate the amount of waste by the load capacity of a truck. "Jūryō-tonne [重量トン]", "weight-tonne", estimated the quantity by weighing a truck loaded with waste. "Jitsuryō-tonne [実量トン]", "actual weight-tonne", measured the

weight of waste by cranes in a plant. The opposition was against "Seisō-tonne" in that it overestimated the amount (Meguro Seisō Kōjō no Kensetsu ni Hantai Suru Jimoto Yūshi no Kai, 1993).

58 Gross prefectural expenditure at constant prices based on Prefectural Accounts 1975–1999 (68SNA, benchmark year=1990).

59 Tōkyō-to Gyōzaisei Kinkyū Taisaku Purojekuto Chīmu [東京都行財政緊急対策プロジェクトチーム] (Tōkyō-to Kikaku Chōsei Shitsu, 1975).

60 Tōkyō-to Gyōzaisei Kaizen Iinkai [東京都行財政改善委員会] (Tōkyō-to Gyōzaisei Kaizen Iinkai, 1975).

61 Toseido Chōsa Kai [都制度調査会] (Toseido Chōsa Kai, 1984).

62 Tokubetsu-ku-sei Chōsakai [特別区政調査会].

63 Toku Seido Kaikaku no Kihon Hōkō [都区度改革の基本方向] (Toku Seido Kentō Iinkai, 1986).

64 In 1981, the TMG and the wards launched a joint research body to study how to manage collection and transportation if devolved to the wards (Seisō Jigyō Ikan Mondai Kyōgikai, 1984).

65 A local government association (Ichibu Jimu Kumiai [一部事務組合]) consists of a number of municipalities and jointly conducts administrative services which are too difficult or too inefficient to be carried out by a single municipal government. It is often formed for fire services, waste disposal, cremation and so forth.

66 Chihō Seido Chōsa Kai [地方制度調査会].

References

Arakawa-ku (1989). *Arakawa-ku-shi: Gekan.*

Asahi Shimbun. (1971). Seisō Kōjō Hantai Undō sono Ronri. *Asahi Shimbun Sha*, 7.10.1971.

Asahi Shimbun. (1972a). Shōgatsu Gomi Atsume Kiki. *Asahi Shimbun Sha*, 16.12.1972.

Asahi Shimbun. (1972b). Suginami Kōjō Raigetsu Hatsuka made ni Medo. *Asahi Shimbun Sha*, 28.1.1972.

Asahi Shimbun. (1973a). Kensetsu Sokushin Ha Ugokidasu. *Asahi Shimbun Sha*, 18.9.1973.

Asahi Shimbun. (1973b). Kōsō Biru Shōkyakujō Kōsō no Hankyō. *Asahi Shimbun Sha*, 12.9.1973.

Asahi Shimbun. (1973c). Shinjuku Seisō Kōjō Kōsō Jimoto kara Fukurodataki. *Asahi Shimbun Sha*, 14.9.1973.

Asahi Shimbun. (1974). To no Seisō Kōjō Kensetsu Pēsu Ōhaba Daun. *Asahi Shimbun Sha*, 21.9.1974.

Asahi Shimbun. (1976a). Arakawaku ni Gesuishorijō. *Asahi Shimbun Sha*, 25.2.1976.

Asahi Shimbun. (1976b). Keimusho Atochi wa Bōsai Hiroba ni. *Asahi Shimbun Sha*, 28.11.1976.

Asahi Shimbun. (1976c). Shinjuku Gomi Biru Kōsō Shibomu. *Asahi Shimbun Sha*, 23.1.1976.

Asahi Shimbun. (1976d). Shinjuku Gomi Biru Kōsō Tanaage. *Asahi Shimbun Sha*, 24.1.1976.

Asahi Shimbun. (1977). Seisō Kōjō no Kawari ni Toritsu Kō. *Asahi Shimbun Sha*, 23.1.1977.

Asahi Shimbun. (1980). Gomi Jikunaishori no Moderu Shibuya Seisō Kōjō Chū ni. *Asahi Shimbun Sha*, 31.1.1980.

Haikibutsu Gakkai (ed.). (2003). *Gomi Dokuhon Shimban.* Tokyo: Chūō Hōki Shuppan.

Jinno, N. (1995). Toshi Keiei no Hatan kara Saiken e. In N. Jinno (ed.), *Toshi o Keiei Suru* (pp. 67–138). Tokyo: Toshi Shuppan.

Kishimoto, K., & Yorimoto, K. (1976). Kono Hito to Gomi Ron. *Gekkan Haikibutsu*, 34–42.

Kita-ku-gikai. (1994). *Kita-ku-gikai-shi: Tsūshi Hen.*

Kodama, T., & Yokoyama, K. (1974). Tōkyō Gomi Sensō to Shōhi Seikatsu. *Shimin*, 62–74.

Komatsuzaki, G., Shibata, T., & Mishiba, K. (1971). Seisō Mondai to Tōkyō Tosei no Kadai. *Tosei*, 16(11), 6–18.

Komine, T. (ed.). (2011). *Nihon Keizai no Kiroku: Dai Niji Sekiyu Kiki eno Taiō kara Baburu Hōkai made.* Tokyo: Naikakufu Keizai Shakai Sōgō Kenkyūjo.

Kōtō-ku. (1975). Gesui Shorijō Odei Shōkyaku Zampai Chūō Bōhatei Uchigawa Shobunjō Tōki Mondai ni tsuite.

Kōtō-ku. (1977). Owatte inai Gomi Sensō. *Kōtō-Ku-Hō Dai 481 Gō.*

Kōtō-ku-gikai. (1983). Kōtō-ku-gikai Kaigiroku Shōwa Gojūhachinen Dai Nikai Teireikai.

Mainichi Shimbun. (1971a). Kugikai no Dōi Matsu. *Mainichi Shimbun Sha*, 7.12.1971.

Mainichi Shimbun. (1971b). Seisō Kōjō Kensetsu: Ii Tochi Arimasu. *Mainichi Shimbun Sha*, 8.12.1971.

Mainichi Shimbun. (1972). Toneri wa Naze Dame?. *Mainichi Shimbun Sha*, 13.4.1972.

Mainichi Shimbun. (1973a). Fukutoshinkaihatsu-kyō Zenin ga Hantai. *Mainichi Shimbun Sha*, 14.9.1973.

Mainichi Shimbun. (1973b). Shinjuku Seisō Kōjō Shinkoku na Kabe ni. *Mainichi Shimbun Sha*, 18.9.1973.

Mainichi Shimbun. (1973c). Suginami Seisō Kōjō Nankō: Kōtō-ku nimo Kōnan Ryōron. *Mainichi Shimbun Sha*, 30.9.1973.

Mainichi Shimbun. (1974). Gomi Taii Shukushō shi Saisei. *Mainichi Shimbun Sha*, 10.5.1974.

Mainichi Shimbun. (1976). Dannen Shimasu Tokonoma Gomi Kōjō. *Mainichi Shimbun Sha*, 10.7.1976.

Mainichi Shimbun. (1977a). Arakawa-ku no Gesuidō Sensō Ketchaku e. *Mainichi Shimbun Sha*, 7.12.1977.

Mainichi Shimbun. (1977b). Kōtō Gomi Sensō yatto Ketchaku. *Mainichi Shimbun Sha*, 25.5.1977.

Mainichi Shimbun. (1977c). Shūshūan ni Kihon Gōi. *Mainichi Shimbun Sha*, 27.5.1977.

Meguro Seisō Kōjō no Kensetsu ni Hantai Suru Jimoto Yūshi no Kai. (1993). *Furimukeba Entotsu.*

Mochida, N. (1995). Shuto Zaisei no Keizai Gaku. In N. Jinno (ed.), *Toshi o Keiei Suru* (pp. 13–66). Tokyo: Toshi Shuppan.

Naito, Y. (2005). *Takaido no Konjaku to Tōkyō Gomi Sensō.* Tokyo: Akarui Seikatsu Sha.

Nakano Eki Shūhen Chiku Seibibu. (1991). Koremade no Omo na Keika.

Nakasugi, O. (1982). Gomi Shori Shisetsu Kensetsu no Hiyō Bunseki to Kongo no Kadai. In K. Yamamoto (ed.), *Gendai no Gomi Mondai* (pp. 143–163). Tokyo: Chūō Hōki Shuppan.

Nakazawa, T. (2017). What Is against an Idea of Distributive Justice? Local Responses to In-Ward Waste Disposal in Tokyo. *Environmental Sociology*, 3(3), 213–225.

Nerima-ku-gikai. (1991). *Nerima-ku-gikai-shi.*

Osumi, S. (1972). *Gomi Sensō.* Tokyo: Gakuyō Shobō.

Ōta-ku-gikai. (1973). Ōta-ku-gikai Kaigiroku: Shōwa Yonjūhachinen Dai Sankai Teireikai.

Ōta-ku-gikai. (2003). *Ōta-ku-gikai-shi.*

Seisō Jigyō Ikan Mondai Kyōgikai. (1984). *Seisō Jigyō Ikan Mondai Kyōgikai Hōkoku.*

Shakai Chōsa Kenkyūjo. (1973). *Tōkyō no Gomi Sensō to Tomin no Kanshindo.*

Shibata, Tokue. (2006). Land, Waste & Pollution, In H. Tamagawa (ed.), *Sustainable Cities,* Tokyo: United Nations University Press.

Shibata, Teruyoshi. (2001a). Seijiteki Funsō Katei ni okeru Masu Media no Kinō 1. *Hokudai Hōgaku Ronshū,* 51(6), 1929–1959.

Shibata, Teruyoshi. (2001b). Seijiteki Funsō Katei ni okeru Masu Media no Kinō 2. *Hokudai Hōgaku Ronshū,* 52(2), 573–601.

Shibuya-ku-gikai. (1976). *Shibuya-ku-gikai-shi.*

Shōyō Kinen Zaidan. (1983). *Tōkyō Gomi Sensō: Takaido Jūmin no Kiroku.*

Toku Seido Kentō Iinkai. (1986). *Toku Seido Kaikaku no Kihon Hōkō.*

Tokubetsu-ku-sei Chōsakai. (1981). *Tokureishi no Kōsō.*

Tōkyō Nijūsan-ku Seisō Ichibu Jimu Kumiai. (2006). *Ippan Haikibutsu Shori Kihon Keikaku.*

Tōkyō Nijūsan-ku Seisō Ichibu Jimu Kumiai. (2013). *Jigyō Gaiyō Heisei Nijūgonendo Ban.*

Tōkyō Seisō Rōdōkumiai. (1981). *Tōkyō Seisō Rōdōkumiai Sanjūnen-shi.*

Tōkyō-to. (1971). *Gomi Mondai ni Kansuru Seron Chōsa Hōkokusho.*

Tōkyō-to. (1973). *Gomi Mondai ni Kansuru Seron Chōsa.*

Tōkyō-to. (1976a). Chōgi Yōroku: Dai Sankai.

Tōkyō-to. (1976b). *Tōkyō-to no Gyōzaisei Sankanen Keikaku.*

Tōkyō-to. (1982). *Tōkyō-to Chōki Keikaku: Mai Taun Tōkyō 21 Seiki o Mezashite.*

Tōkyō-to. (1994). *Tōkyō-to-sei Gojūnen-shi.*

Tōkyō-to Gomi Taisaku Hombu. (1971). Gomi Sensō Mittsu no Honshitsu to Itsutsu no Mondai. *Gomi Sensō Shūhō No.1.*

Tōkyō-to Gomi Taisaku Hombu. (1972a). Adachi-ku Seisō Kōjō Tatekae ni tsuite Yōbō. *Gomi Sensō Shūhō No.48.*

Tōkyō-to Gomi Taisaku Hombu. (1972b). Suginami Kumin ni Gomi Genryō no Yobikake. *Gomi Sensō Shūhō No.53.*

Tōkyō-to Gomi Taisaku Hombu. (1973). Kaku Chiku no Ugoki. *Gomi Sensō Shūhō No.54.*

Tōkyō-to Gomi Taisaku Hombu. (1974). Gomi Mondai no Naka nimo Tosei no Jishin o Sasaeru Yōso. *Gomi Sensō Shūhō No.104.*

Tōkyō-to Gyōzaisei Kaizen Iinkai. (1975). *Atarashii To to Tokubetsu-ku no Kankei Kakuritsu no tame no Iken.*

Tōkyō-to Kikaku Chōsei Shitsu. (1971). *Tōkyō-to Chūki Keikaku 1970.*

Tōkyō-to Kikaku Chōsei Shitsu. (1972). *Tōkyō-to Chūki Keikaku 1971.*

Tōkyō-to Kikaku Chōsei Shitsu. (1975). *Tōkyō-to Gyōzaisei Kinkyū Taisaku Purojekuto Chīmu Hōkokusho.*

Tōkyō-to Kikaku Hōdō Shitsu. (1981). *Mai Taun Tōkyō 81.*

Tōkyō-to Kikaku Shingi Shitsu. (1986). *Dai Niji Tōkyō-to Chōki Keikaku.*

Tōkyō-to Seisaku Shitsu. (1978). *Teiseichō Shakai to Tosei.*

Tōkyō-to Seisō Kyoku. (1971). *Tōkyō no Gomi.*

Tōkyō-to Seisō Kyoku. (1977). *Jigyō Gaiyō.*

Tōkyō-to Seisō Kyoku. (1992). *Tōkyō-to Meguro Seisō Kōjō Kensetsu no Enkaku to Keii.*

Tōkyō-to Seisō Kyoku. (1994). *Gomi Mondai Kinkyū Repōto: Dai Yonkai.*

Tōkyō-to Seisō Kyoku. (1995). *Tōkyō Gomi Hakusho.*

Tōkyō-to Seisō Kyoku. (2000). *Tōkyō-to Seisō Jigyō Hyakunen-shi.* Tokyo: Tōkyō-to Kankyō Seibi Kōsha.

Tōkyō-to Seisō Shingikai. (1976). *Gomi Ryō no Henka ni Tomonau Seisō Shisetsu Seibi no Kihonteki Kangaekata ni tsuite.*

Tōkyō-to-gikai. (1972a). Tōkyō-to-gikai Eisei Keizai Seisō Iinkai Sokkiroku: Shōwa Yonjūnananen Dai Nigō.

Tōkyō-to-gikai. (1972b). Tōkyō-to-gikai Eisei Keizai Seisō Iinkai Sokkiroku: Shōwa Yonjūnananen Dai Nijūsangō.

Tōkyō-to-gikai. (1973). Tōkyō-to-gikai Kaigiroku Shōwa Yonjūhachinen Dai Yonkai Teireikai.

Tōkyō-to-gikai. (1974a). Tōkyō-to-gikai Yosan Tokubetsu Iinkai Sokkiroku: Shōwa Yonjūkyūnen Dai Gogō.

Tōkyō-to-gikai. (1974b). Tōkyō-to-gikai Yosan Tokubetsu Iinkai Sokkiroku: Shōwa Yonjūkyūnen Dai Yongō.

Tōkyō-to-gikai. (1975a). Tōkyō-to-gikai Kaigiroku Shōwa Gojūnen Dai Ikkai Teireikai.

Tōkyō-to-gikai. (1975b). Tōkyō-to-gikai Kaigiroku Shōwa Gojūnen Dai Yonkai Teireikai.

Tōkyō-to-gikai. (1976a). Tōkyō-to-gikai Eisei Keizai Bukka Seisō Iinkai Sokkiroku: Shōwa Gojūichinen Dai Jūichigō.

Tōkyō-to-gikai. (1976b). Tōkyō-to-gikai Yosan Tokubetsu Iinkai Sokkiroku: Shōwa Gojūichinen Dai Yongō.

Tōkyō-to-gikai. (1978). Tōkyō-to-gikai Eisei Keizai Seisō Iinkai Sokkiroku: Shōwa Gojūsannen Dai Nanagō.

Tōkyō-to-gikai. (1982). Tōkyō-to-gikai Kensetsu Seisō Iinkai Sokkiroku: Shōwa Gojūnananen Dai Nijūsangō.

Toseido Chōsa Kai. (1984). *Atarashii Toseido no Arikata*.

Tsugawa, T. (1993). *Dokyumento Gomi Kōjō*. Tokyo: Gijutsu to Ningen.

Yomiuri Shimbun. (1971). Kōtō no Shitsumonjō ni Jūnanaku ga Kaitō. *Yomiuri Shimbun Sha*, 12.10.1971.

Yomiuri Shimbun. (1973). Tōkyō Gomi Sensō mata Kimpaku. *Yomiuri Shimbun Sha*, 1.10.1973.

Yorimoto, K. (1977). Sanka to Chiiki Seiji o meguru Gomi no Seijigaku. *Chiiki Kaihatsu*, 4, 1–57.

5   The second garbage war and the revival of IWWD

Despite its weak influence in the second period, In-Ward Waste Disposal (IWWD) regained prominence in the 1990s. As explained in the previous chapters, the incinerator siting of Tokyo had been based on a scheme which intended to locate incinerators in the suburb area along the belt lines and at the coastal reclaimed lands; the central part had been excluded from the siting of incinerators. Although IWWD changed this siting scheme in the first garbage war, the incinerator siting plan went back to the old scheme as IWWD lost its prominence. However, IWWD revived and became influential once again in the 1990s. The incinerator siting regained its momentum to go beyond the old scheme once again. Moreover, the idea of the self-responsibility of each ward in waste disposal, which had not been reflected in the policies even in the early 1970s, became influential as well. The two components of IWWD started working together; the influence of the idea of distributive justice reached its climax in the 1990s.

This chapter looks into what revitalised the influence of this idea of distributive justice in this period. The sudden growth of waste in the latter half of the 1980s as a result of the bubble economy prepared favourable environments for the revival of the idea of siting incinerators in every ward. Furthermore, the politics of the autonomy system reform in the 23 wards generated a driving force for the self-responsibility of each ward in waste disposal, which led to further advancement of the One Ward One Incinerator (OWOI) policy.

The culmination of IWWD's influence in the 1990s

Revival of One Ward One Incinerator policy

The influence of OWOI on siting policies revived in the 1990s. The turning point came when the Tokyo Waste Management Advisory Committee issued a report in 1990, which recommended that new incinerators with a capacity of disposing of 3,400 tonnes/day should be built by 2000 and the siting of them should be based on IWWD (Tōkyō-to Seisō Shingikai, 1990b). The report endorsed IWWD and stated that this idea would contribute to the responsibility of emitters, the fairness among communities, and efficiency in garbage transportation. While the difficulty in finding land for incinerators in the hyper-urbanised area was acknowledged, the report argued that the wards having no incinerator should make an effort to secure land, even if it was only suitable for a small plant. It also required prioritising incinerator construction in the use of land owned by the TMG, the sites of national institutions and private companies, reclaimed land in the coastal area, and large-scale development projects.

IWWD was substantively reflected in the Incinerator Construction Plan[1] in 1991. Receiving the report, the TMG announced its incinerator siting plan in October 1991, which projected to construct ten new incinerators by 2011. This 20-year plan emphasised IWWD as a fundamental principle in the siting of incinerators and basically targeted wards without incinerators. As shown in Figure 5.1, there were 11 wards left without an incinerator back then.[2] Although two of the ten projects (the third incinerator in Setagaya and the incinerator vessel to be harboured at the current landfill next to Koto) were considered as an emergency measure to achieve All Waste Incineration (AWI) as soon as possible[3] and planned in the wards in which incinerators were sited, the other eight projects were located in wards without an incinerator: Chiyoda, Chuo, Minato, Sumida, Shibuya, Nakano, Toshima, and Arakawa. Twenty wards out of the 23 would have at least one incinerator by 2011, if the projects proceeded as planned. While no concrete plan was drawn for the other three wards, i.e. Shinjuku, Bunkyo and Taito, the TMG promised that they would be kept in view in case of unpredictable

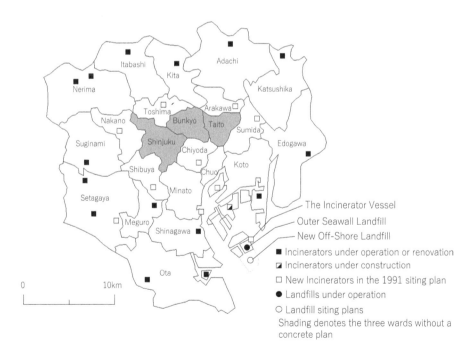

Figure 5.1   Location of incinerators and landfills in 1991.

Adapted from a map of Tokyo by CraftMap (URL: http://www.craftmap.box-i.net/) based on Tōkyō-to Seisō Kyoku (1991c).

Table 5.1 Schedule of incinerators construction in 1991 and 1994

| Projects | 1991 siting plan (1994 advancement) | |
	Construction	Operation
Sumida	1993	1996
Setagaya	1993	1996
Incinerator vessel	1993	1996
Minato	1994	1998
Toshima	1994	1998
Shibuya	1994	1998
Chuo	1997	2001 (2000)
Chiyoda	1997	2002
Nakano	2004 (2002)	2008 (2006)
Arakawa	2007 (2002)	2011 (2006)
Shinjuku	not specified (2002)	not specified (2006)
Taito	not specified (searching for a site)	
Bunkyo	not specified (searching for a site)	

Source: Toku Kyōgikai & Toku Seido Kaikaku Suishin Iinkai (1994), Toku Seido Kaikaku Suishin Iinkai (1994), Tōkyō Seisō Rōdōkumiai & Tōkyō-to (1994), and Tōkyō-to Seisō Kyoku (1991c).

socio-economic changes in the future. Thus, IWWD substantially influenced the siting of incinerators in the 1991 plan.

The incinerator siting was further accelerated through the first half of the 1990s (Table 5.1). In 1994, the candidate site for the project in Shinjuku, which was not specified in the 1991 siting plan, was announced. The Shinjuku plant was scheduled to start construction in 2002 and operation in 2006. On top of that, the TMG announced the advancement of the schedule for the projects in Nakano and Arakawa, which were originally planned to start construction in 2004 and in 2007 respectively, to the same schedule as Shinjuku (Tosei Shimpō, 1994a). The Chuo plant was also to become operational one year earlier than what was stated in the 1991 siting plan. This acceleration was closely related to the development of the political situation around the autonomy expansion movement.

Rising influence of IWWD in devolution policy

IWWD became a dominant idea in the devolution of waste management as well. As noted earlier, the autonomy of the 23 wards had been limited compared with normal local municipalities under the Special Wards System. The 23 wards had initiated the autonomy expansion movement in 1947. The devolution of waste management services had been one of the most significant, and the most controversial, agenda in this political movement. However, the propositions and arguments over the devolution had been limited to collection and transportation of waste; the idea that each ward should be institutionally responsible for waste disposal was hardly influential throughout the 1970s and the 1980s. Nonetheless, in the 1990s, IWWD became the centre of focus in the devolution and incorporated into policies on institutional reform.

IWWD became a basic principle of the devolution policy. In 1994, the TMG and the 23 wards came to an agreement and prescribed the basic policy in the Outline of Tokyo-Wards System Reform in 1994 (the 1994 agreement).[4] In this agreement, IWWD was defined as an idea that a ward should be responsible for and self-sufficiently perform all of the waste management services in the ward, from collection, transportation, intermediate treatment (including incineration and others), to final disposal (landfilling). The 1994 agreement scheduled to devolve the legal responsibility for waste management to each ward in 2000. The regional disposal approach was to be replaced by the self-sufficient incineration system by each ward via an intermediate system, i.e. the block incineration system in which the 23 wards would be divided into several blocks and incineration would be performed jointly within a block until enough incinerators for the self-sufficient incineration system were constructed. The roadmap to the self-sufficient incineration system was explicitly reflected in the policy for the first time.

Koto's campaign for IWWD

Landfill crisis and Koto's claim for IWWD

The revival of the strength of IWWD was triggered by the protest of Koto ward to the siting of a new landfill in Tokyo Bay. Koto started advocating IWWD again when the ward was asked to accept a new landfill next to the outer seawall landfill which had been operational since 1977 (Figure 3.3). As explained in the previous chapter, the bubble economy in the latter half of the 1980s stimulated waste production, which had remained almost constant at around 3.8 million tonnes/year during the early 1980s. Waste generation had grown by 5.6% every year on average since 1985 and reached 4.9 million tonnes in 1989. This drastic growth of waste brought about the landfill crisis once again in the early 1990s.

The outer seawall landfill, which Koto agreed to in 1977 in exchange for the lands owned by the TMG, was estimated to reach its capacity earlier than expected, given this increased amount of waste (Tōkyō-to Seisō Shingikai, 1990a). In 1981, the landfill was estimated to last only until 1985 (Tōkyō-to Kikaku Hōdō Shitsu, 1981), but it was extended to 1990 in 1982 (Tōkyō-to, 1982) and to 1995 in 1987 (Tōkyō-to Kikaku Shingi Shitsu, 1987), reflecting the low waste growth during the early half of the 1980s. However, the 1990 interim report by the Tokyo Waste Management Advisory Committee anticipated that the landfill would be full in two years and ten months (Tōkyō-to Seisō Shingikai, 1990a).[5] Waste management in Tokyo was falling into a crisis again. As the day of the depletion of the existing landfill was approaching, the TMG had to construct a new landfill as soon as possible.

Facing the landfill crisis, the TMG consulted the Harbour Advisory Committee[6] about a new landfill in 1989. When the committee started examining three candidate sites,[7] Koto ward sent an opinion letter to the governor of Tokyo, which demanded close consultation with the ward on site selection and requested considering the fair distribution of the waste disposal burden among the 23 wards

(Kōtō-ku-gikai, 1989a). Nonetheless, in early 1991, the committee submitted an interim report which recommended the outside area of the current landfill (Tōkyō-to Kōwan Shingikai Kaimen Shobunjō Kentō Bukai, 1991a). This decision angered the ward and brought about the second garbage war. At the end of the first garbage war, Koto accepted the outer seawall landfill in 1977 after a long negotiation with the TMG. The ward complained that siting the new landfill meant that people of the ward would have to continue suffering the burden of waste disposal.

The ward immediately launched a campaign against the new landfill. Koto demanded the TMG to keep the promises it made with the ward in the first garbage war, i.e. IWWD and dispersed dumping.[8] As argued in the previous chapters, although IWWD was adopted by the TMG as a fundamental principle in waste management policies, this idea of distributive justice had hardly been realised and lost its influence in the policies during the 1980s. Koto complained that the number of wards with incinerators increased by only two, Meguro and Suginami, in the past 20 years, leaving the 11 wards without even plans to site an incinerator. Furthermore, AWI, the goal for which the ward championed IWWD, had not been achieved yet; around 20% of the waste for incineration was not incinerated and sent directly to the landfill. Given the situation, Koto launched a protest campaign against the new landfill. In this campaign, IWWD once again became the slogan to highlight the disproportionate burden that Koto suffered. In other words, the ward's interest in this idea of distributive justice, which once declined in the second period, grew in influence due to the impact of the bubble economy and the resultant landfill crisis.

The political veto over landfill and garbage blockade

In this campaign, the ward again took advantage of the de facto veto power over the new landfill. To site a landfill in the sea, it was necessary to hear the opinions of the mayors of the concerned local governments in the process of environmental impact assessment.[9] The process of issuing the license for landfilling in public waters also involved the hearing of mayors' opinions based on resolutions by the local councils.[10] Although the consent of the concerned wards was not a legal requirement, their opinions were one of the significant factors in the project approval process. For the TMG, the new landfill was the most urgent matter and any delay of the project could have led to a serious garbage crisis. In this situation, this opinion hearing in the new landfill siting process effectively served as a bargaining chip for Koto ward to exert its influence on the policy-making process.

Furthermore, Koto threatened the TMG and other wards to block waste being sent to the current landfill as it did in the first garbage war. Remembering what happened in Suginami in the first garbage war, the prospect of a blockade was frightening not just for the TMG but also for the rest of the 23 wards, and, in particular, for the 11 wards without any incinerators.[11] Thus, by taking advantage of the decision-making process of the landfill siting and the threat of a blockade, Koto was able to make the TMG and the other wards take its concerns seriously.

Effectively, the 1991 siting plan was made as a result of this pressure from Koto. In the protest against the new landfill, Koto ward criticised the TMG for its lack of a long-term perspective on waste management and the tendency to easily rely on landfill disposal (Kōtō-ku-gikai, 1989b, 1989c). When the Harbour Advisory Committee was about to submit the interim report, the mayor of Koto and the members of the Tokyo Metropolitan Assembly from the ward insisted that Koto ward would not agree on the new landfill unless a long-term vision on waste management was shown (Kōtō-ku-gikai, 1991b).[12] In order to persuade the ward, the TMG started making long-term waste management schemes, one of which resulted in the 1991 siting plan showing a 20-year prospect with ten new incinerators scheduled to be built by 2011.[13]

Koto continued pressuring the TMG and the other wards. Before the 1991 plan was made public, Koto visited the other 23 wards and handed to the mayors and the chairpersons a letter which complained that IWWD was not being achieved and asked for their cooperation on the siting of incinerators in their jurisdiction (Kōtō-ku-gikai, 1991a; Tosei Shimpō, 1991b). The de facto veto over the new landfill was utilised to influence the policy making. When the Harbour Advisory Committee was about to submit the final report on the new landfill in August 1991, the mayor of Koto stated in the committee that the new landfill would not be constructed if he refused, as his opinion must be heard in the process of issuing the license for the landfill (Kōtō-ku-gikai, 1991b). As a result, the final report was suspended until Koto was fully consulted on the issue (Kōtō-ku-gikai, 1992a), which troubled the TMG as the capacity of the existing landfill would soon be reached. In August 1991, officers of the Koto government and all of the council members went to the TMG and threatened it with a blockade of the garbage trucks (Asahi Shimbun, 1991a, 1991b). It was under this pressure from Koto that the incinerator siting plan was created.

Koto's campaign continued even after the 1991 siting plan was announced. Not satisfied with this plan, the ward sent to the TMG another opinion letter, which demanded the following: the cancellation of the incinerator vessel to be harboured at the current landfill; incinerator projects for the three wards without concrete plans; and a guarantee that incinerators would be built as planned in the eight wards (Kōtō-ku-gikai, 1991d). Dissatisfied with the ambiguous reply from the TMG, the ward refused to hold the Harbour Advisory Committee to issue the final report on the new landfill, thereby further delaying the siting process of the new landfill.[14]

On top of that, Koto warned the TMG of the potential blocking of waste from Shinjuku ward and Ginza in Chuo ward at the end of November 1991 (Kōtō-ku-gikai, 1991e). Shinjuku was targeted not just because no concrete plan was shown for the ward in the 1991 siting plan, but also because the siting plan in Shinjuku during the first garbage war was cancelled in part due to local opposition, even though the ward had one of the busiest commercial streets which produced huge amounts of food waste. Ginza, which was in Chuo ward but close to the incinerator project in Chiyoda ward, became the target because the shopkeepers' association and businesses there had strongly opposed the siting project in Chiyoda arguing that it was not acceptable to have an incinerator at the entrance of Tokyo, next to the

Tokyo Rail Station (Asahi Shimbun, 1991c). Koto contended that the opposition was not tolerable for those who had long suffered garbage trucks driving through the ward, and announced the blockade to make people in Chuo ward realise that their waste was taken care of by the sacrifice of Koto ward (Kōtō-ku-gikai, 1991e).

The Harbour Advisory Committee meeting was finally held, and the blockade was cancelled after Governor Suzuki visited Koto and apologised for the disproportionate burdens imposed on the ward. He promised to firmly proceed with the 1991 incinerator siting plan and to make the utmost effort to construct incinerators in the three wards for which a concrete project was not shown in the plan. At the end of 1991, the Harbour Advisory Committee submitted the final report which recommended that the new landfill was to be sited outside of the outer seawall area, but on condition that the concerned wards, i.e. Chuo, Minato, Koto, Shinagawa, and Ota, be fully consulted over the issue. The siting process of the new landfill started moving forward, but was yet to be approved by Koto as well as the other four wards.

Koto continued to call for the realisation of IWWD and suspended the agreement on the new landfill siting. When consulted over the landfill in April 1992, the ward demanded the fulfilment of the seven requests, which included making concrete plans for the three wards without an incinerator building project, constructing the incinerators in the eight wards as planned, and cancelling the incinerator vessel (Kōtō-ku-gikai, 1992b).[15] It was not until the end of 1995 that Koto finally agreed on the new landfill when the governor[16] visited the ward and apologised for the burden imposed on the ward for years, manifesting his determination to solve the garbage problems. The TMG promised to construct incinerators, even if only small ones, in the remaining two wards for which an incinerator construction plan was not made yet, i.e. Bunkyo and Taito.[17] In addition, the government vowed to reconsider the incinerator vessel project and to offer a block-by-block negotiation of the new landfill, by which the ward would be consulted every time a new block of the landfill would start operation.[18] Koto ward agreed on the new landfill in general "taking into consideration the well-being of the whole citizenry of Tokyo",[19] while maintaining that it was "a tough decision", as accepting the new landfill meant that the sacrifice of the people in the southern part of the ward had to continue for the foreseeable future (Kōtō-ku-gikai, 1996).

Thus, the campaign by Koto played a major role in increasing the influence of IWWD on the 1991 siting plan. Koto took advantage of the new landfill project and the garbage blockade to influence the policy making. The persistent protest by the ward made the TMG and the other wards commit to IWWD and accelerate the siting projects.

Readoption by TMG

Restored consistency with AWI

In the meantime, IWWD became appealing to the TMG. The idea was recognised as a solution to the garbage crisis once again, for the increased amount of waste and the landfill crisis caused by the bubble economy produced an urgent necessity to construct more incinerators. IWWD was readopted by the TMG, not only because of the interest and power of Koto ward, but also due to the restored cognitive legitimacy of the idea that incinerators be sited in every ward.

The TMG advocated IWWD again to achieve AWI, which had been one of its primary policy goals in waste management. In the first garbage war, the TMG made use of IWWD to facilitate the incinerator siting for this policy goal, but abandoned this idea when it thought that AWI could be achieved without constructing as many incinerators as IWWD required. The Tokyo Long-term Plan in 1982 estimated that AWI would be achieved by 1995 with the construction of only five new incinerators in addition to the renovation of the old ones. The incinerator siting had gone back to the old scheme in which siting incinerators in the central area of Tokyo was considered unnecessary, let alone constructing one in every ward. However, the drastic increase of waste during the latter half of the 1980s made it impossible to achieve AWI with this siting scheme. The amount of waste reached 4.9 million tonnes/year in 1989, while the Second Long-term Plan in 1986 estimated that the annual waste generation would remain at around 4 million tonnes/year until 2000. To achieve AWI, the TMG could no longer afford to leave the central area exempted from incinerators.

The immediate need for more incinerators was exacerbated by the deepening landfill crisis. As noted above, the sudden increase of waste resulted in the landfill capacity being consumed more quickly than expected. As the landfill was the final destination of the waste, the depletion of it meant the failure of the entire waste management system. In the history of waste management in Tokyo, AWI had been held as a primary policy goal because it could reduce the volume of garbage entering the landfills. However, given the sudden increase of waste, around 20% of waste for incineration was still not being incinerated and brought directly to the landfills. The TMG had to achieve AWI as soon as possible to save the landfills and avoid the disruption of its waste management system.

Furthermore, the completion of AWI became a prerequisite for the new landfill siting plan. Facing the increased waste and the landfill crisis, the TMG consulted with the Tokyo Waste Management Advisory Committee. The committee's interim report in 1990 stated that untreated waste, both waste for incineration and not for incineration, should not be allowed in the new, and probably the last, landfill. The Long-Term Vision of Waste Management,[20] which was one of the three policies put forward as a response to Koto's request for a long-term vision, also stated that untreated waste would not be acceptable in the new landfill. This provision was incorporated into the final report of the Harbour Advisory Committee as well (Tōkyō-to Kōwan Shingikai Kaimen Shobunjō Kentō Bukai, 1991b). Achieving AWI before the new landfill started operation became a pressing policy goal.

Moreover, learning a lesson from the disruptive increase of waste caused by the booming economy, AWI was required in a stricter sense to unfailingly incinerate all waste for incineration in the future. Having 30% margins was recommended to cope with the seasonal fluctuation of the amount of waste as well as troubles, regular inspections, and overhauls of incinerators (Tōkyō-to Kōwan Shingikai Kaimen Shobunjō Kentō Bukai, 1991a, 1991b; Tōkyō-to Seisō Shingikai, 1990b). This strict requirement, which was called the stable incineration system, necessitated more incinerators than AWI originally required.

Thus, IWWD regained its prominence as a problem-solving idea due to the sudden increase of waste and the landfill crisis. The TMG considered that AWI could not be achieved without constructing as many incinerators as this idea of distributive justice required. The rise of IWWD during this period was supported not only by the interests and power of Koto ward, but also by its revived cognitive legitimacy.

The bursting of the bubble and its consequent impact

It is noteworthy that the bubble economy had already ended by the time the 1991 siting plan was announced. The bubble economy, which boomed in the late 1980s, burst in 1989 and Japan had entered a long recession. To cool down the overheated economy, the Bank of Japan implemented a monetary tightening policy of raising the official interest rate. The transaction volume of assets was regulated to restrain property prices. These policies put an end to the economic boom. The Nikkei Stock Average crashed below 20,000 yen in 1990, after it hit a record high of 38,915 yen at the end of 1989. In Tokyo, the economic growth rate sharply dropped to −0.1% on average during the 1990 to 1995 period.[21] Given this sudden economic change, the amount of garbage also started decreasing after reaching a peak of 4.9 million tonnes/year in 1989 (Tōkyō Nijūsan-ku Seisō Ichibu Jimu Kumiai, 2006; Tōkyō-to Seisō Kyoku, 2000).

Nonetheless, this economic downturn and the resultant decrease in waste did not affect the siting policies so much during the early half of the 1990s. This was in part because the TMG estimated that the quantity of waste would increase in the next 25 years. Although the actual amount of waste started decreasing after 1989, the 1991 siting plan was still based on the estimation that the amount would slightly increase from 4.81 million tonnes/year in 1990 to 5.21 million tonnes/year in 2015. Given this estimation, more incinerators were believed to be necessary to achieve AWI. The effect of the economic change on the expectations of policy makers was not strong enough to invalidate the idea that incinerators should be sited in every ward.

Similarly, although the bursting of the bubble economy started aggravating the financial condition of Tokyo, it had yet to make a considerable impact upon the incinerator siting policy in the early 1990s. The TMG's finance, which deteriorated after the 1973 oil crisis, was reconstructed immediately under the Suzuki administration during the 1980s (Jinno, 1995). In addition, the bubble economy increased the tax revenue dramatically.

It is true that there was concern for the financial feasibility of the 1991 siting plan (Tōkyō-to-gikai, 1991b). The incinerators project was estimated to cost 720 billion yen over a 20-year period (Tōkyō-to Seisō Kyoku, 1991c). Even though the asset bubble burst in 1989, property prices were still extremely high, especially in the centre of Tokyo. The cost of the land purchase in Minato, Shibuya and Toshima was estimated to be 190 billion yen, which would amount to 2.5% of the TMG's total budget in 1993 (Tosei Shimpō, 1991c).

Nonetheless, the implementation of the 1991 siting plan was not in doubt during this period. Although tax revenue started decreasing since the bubble burst, the TMG could maintain the budget level by issuing government bonds and using the huge government reserve funds which were raised in the 1980s in preparation for economic fluctuation (Tōkyō-to Gyōzaisei Kaikaku Suishin Hombu, 1996). The optimistic prospect that the economy would recover also helped this expensive project. When the plan was outlined in 1991, the economy of Tokyo was anticipated to grow by more than 4% annually until 2000 (Tōkyō-to Kikaku Shingi Shitsu, 1990). Furthermore, the project was one of the governor's public promises and believed to be urgent. It was not until 1997 that the deteriorating financial condition led to the revision of the incinerator siting plan.

Sustainable Waste Management

It is also worth mentioning that a new paradigmatic policy idea of waste management was emerging in this period, i.e. Sustainable Waste Management (SWM). This policy paradigm was to extend the scope of waste management to the process of waste production and introduce up-stream measures such as reduction, reuse, and recycling (3Rs), in contrast to incinerationism, which concerned only the processes after waste was produced. Until the early half of the 1980s, waste management had focused on building disposal facilities such as incinerators and landfills to deal with the growing amount of waste. Under incinerationism, constructing more incinerators had been encouraged to catch up with the growth of waste amidst rapid economic growth. However, facing the difficulty in constructing incinerators and the depletion of landfill space, depending solely on "the end of the pipe" measures was increasingly being considered unsustainable. That was when SWM rose to prominence as a new policy paradigm. More emphasis was placed on reducing the amount of waste to be disposed of, as this new policy paradigm became influential in Japan. Actually, from the 1990s to the early 2000s, the Japanese waste management system was drastically changed, and many new pieces of legislation were passed.[22] The TMG also incorporated this idea into its policies and started making systematic efforts to reduce the amount of waste.

This rise of SWM worked negatively against IWWD, which was intimately connected with incinerationism, for this new rising idea undermined the validity of expanding incineration capacity. Firstly, SWM provided an alternative to expanding incineration capacity to solve waste problems. Although incineration was still the dominant technology for waste disposal, it was no longer seen as the only way to solve waste problems. Rather, incineration was relegated to the bottom of the waste management hierarchy under the new policy paradigm. Secondly, SWM became the theoretical backbone for the anti-incineration movements. This idea allowed local opposition and environmental movements to cast doubt on the very need for more incinerators and damaged the cognitive legitimacy of IWWD, which was closely connected with constructing more incinerators as a conventional end-of-pipe policy. Thirdly, the advancement of SWM not only reduced the actual amount of waste, but also lowered the estimation of waste generation for the future. After reaching its peak in 1989, the actual amount of waste kept decreasing due to the advancement of the 3Rs, combined with the impact of the economic recession. Further development of waste reduction methods was set as a policy goal and incorporated into the calculation of the amount of waste generation in the future. Thus, the rise of SMW in waste management gradually made new incinerators less necessary.

In other words, the rise of SWM meant that another idea of distributive justice, i.e. source reduction, was becoming influential. While the advocates of IWWD had taken for granted the necessity of more incinerators and focused on how to distribute them, source reduction would ultimately make having more incinerators unnecessary. This new idea of distributive justice was becoming prevalent among the key actors in the 1990s. The effect of the 3Rs was incorporated in the calculations of the future waste quantity in governmental policies. Opposition movements against the incinerator projects utilised the idea of source reduction to argue against IWWD. The advent of source reduction in siting policies damaged the ideational legitimacy of IWWD and led to its decline.

However, IWWD became influential in the 1990s again despite the rising influence of SMW and the idea of source reduction. This was because the expected impact of SMW on waste reduction was still not enough to reduce the actual necessity for more incinerators. The TMG took into account the effect of waste reduction measures in its estimation of future waste growth in the 1991 siting plan. It was calculated that the potential amount of waste would increase by 2% every year, based on the garbage growth trend in the past 10 years (Tōkyō-to Seisō Kyoku, 1991c; Tōkyō-to Seisō Shingikai, 1990b). Without the source reduction measures, the amount of waste was anticipated to grow to 6.15 million tonnes in 1996, 6.44 million tonnes in 2000 and 7.03 million tonnes in 2015; the waste reduction measures, which aimed to reduce waste by 23% in 2000, were expected to restrain this potential waste increase (Tōkyō-to Seisō Kyoku, 1991c). However, even taking into account this waste reduction effect, the amount was expected to increase to 5.21 million tonnes in 2015. The TMG still believed that more incinerators were necessary to achieve AWI.

Rather, IWWD and SMW were working together to save the precious landfill space. As noted earlier, what made IWWD appealing to the TMG was the urgent need to construct more incinerators to achieve AWI and reduce the amount of waste entering the landfills. Although SMW was becoming influential and incinerationism relatively less influential in waste management policies, the source reduction effort and the construction of more incinerators coexisted in governmental policies during this period, for both were meant to save landfill capacity, given that 20% of waste for incineration was still not being incinerated. Thus, the TMG pushed forward with both the 3Rs and AWI in order to save landfills. It was not until 1997 that the contradiction between SMW and IWWD surfaced.

IWWD in the politics of autonomy reform

The union's strategy on the devolution

Not only the idea of siting incinerators in every ward, but also the self-responsibility of each ward in incineration became influential in this period. As explained in the previous chapters, the intervention of the Tokyo Cleaning Workers' Union had prevented this requirement of IWWD from influencing the waste management policies. In the 1990s, however, the same intervention by the union engendered a political driving force for the idea of self-responsibility of each ward.

The autonomy expansion movement was reaching its climax in the 1990s. Since the amendment of the Local Autonomy Act in 1974, which restored the public election for the ward mayor, the 23 wards had continued this political movement. Expectations for reform rose as Governor Suzuki came into office in 1979, because he publicly committed himself to autonomy reform. Suzuki, as a former administrative vice minister, was influential at the Ministry of Home Affairs[23] which had jurisdiction over the relevant laws. In 1986, the TMG and the 23 wards came to an agreement on the autonomy reform and jointly issued the Basic Policy of Tokyo-Wards System Reform, which aimed at positioning each ward as a normal local municipality, devolving authority and administrative services to each ward, and enhancing its financial autonomy (Toku Seido Kentō Iinkai, 1986). With this agreement, the TMG and the 23 wards petitioned the central government to amend the Local Autonomy Act.

The devolution of waste management became the largest political concern in this movement. To amend the Local Autonomy Act, the TMG and the 23 wards had to persuade the Ministry of Home Affairs (MOH). However, the MOH was reluctant to support the autonomy reform and imposed conditions (Miyake, 2006; Narita, 1998). The 22nd Local Government System Research Council,[24] an advisory committee of the central government, issued a report which agreed that the 23 wards would become basic local municipalities, but on a condition that waste management be devolved in each ward at the same time (Dai Nijūniji Chihō Seido Chōsakai, 1990). The fate of the whole autonomy expansion hinged on the devolution of waste management.

To make matters worse for the TMG and the 23 wards, the Tokyo Cleaning Workers' Union was granted a political veto over the devolution of waste management. The MOH required the consent of the concerned parties over the devolution issue; the TMG and the 23 wards had to make an agreement with the Tokyo Labour Union,[25] of which the Tokyo Cleaning Workers' Union was an influential sub organisation. The cleaning union organised around 10,000 workers in the beginning of the 1990s, which accounted for around 20% of the Tokyo Labour Union (Miyake, 2006). Although the Tokyo Labour Union generally agreed with the autonomy expansion of the 23 wards, it could not accept the devolution of waste management as the cleaning workers had persistently opposed such a development. Thus, to accomplish the entire autonomy reform, the TMG and the 23 wards had to persuade the Tokyo Cleaning Workers' Union.

IWWD was reflected in the policies through the negotiation between the union and the TMG (Tōkyō Seisō Rōdōkumiai, 1999: Tōkyō Tokusyokuin Rōdōkumiai, 1993a, 1993b, 1993c, 1993d, 1995). In the negotiating process, the union claimed IWWD as a basic principle of the devolution. As noted earlier, the union had been firmly opposed to the devolution as it was against their interest. Even the devolution of collection and transportation was not acceptable for the union as these sections had the largest number of workers. It was the union's opposition that had prevented IWWD from becoming influential, as it feared that the idea of the self-responsibility of each ward would facilitate the devolution. Paradoxically, however, the union rather persistently called for IWWD to be embedded in the devolution during the 1990s, although the realisation of this idea was against its organisational interest.

The union insisted that devolving only collection and transportation (the partial devolution) would not enhance the autonomy of the 23 wards. Before the negotiation with the union started, the TMG and the 23 wards intended to devolve only collection and transportation. The wards and the TMG argued that even if it was partial, the devolution would enable each ward to design the collection and transportation by itself in a way uniquely tailored to the ward, thereby making each ward more autonomous. The union argued against this view by insisting that how to perform the collection and transportation was completely dependent on how disposal facilities were operated. Given the shortage and uneven distribution of waste disposal facilities, contended the union, a centralised management of the destination of waste would be necessary, and a ward could not decide how to collect and transport the waste by itself unless disposal was performed self-sufficiently. The union, therefore, insisted that if the 23 wards wanted to become basic local municipalities, each ward should self-sufficiently perform this task and be responsible for not only collection and transportation, but all of the waste management processes including incineration and final disposal.

Behind this claim by the union was its strategic intention to stop the devolution. Given the difficulty to site an incinerator in every ward, self-sufficient incineration by each ward was supposedly unachievable. The union intended to prevent the devolution by making IWWD a prerequisite for this institutional reform, arguing that the devolution should be shelved until a self-sufficient disposal system was established for each ward to be responsible for its own waste management (Toku Kyōgikai & Toku Seido Kaikaku Suishin Iinkai, 1994). In other words, the union challenged the TMG and the wards to realise IWWD first if the wards wanted to become a fully-fledged local autonomy.

IWWD as a principle of the devolution

IWWD was adopted as a fundamental principle of the devolution through the negotiation between the union and the TMG. In April 1993, the union and the TMG came to a common understanding on how waste management should be handled, in which IWWD was adopted as a basic principle (Tōkyō-to, 1993).

In 1994, they issued the Concrete Way of Waste Management,[26] in which IWWD was defined as a principle that required each ward to be responsible for all of the waste management processes and to self-sufficiently perform them.

IWWD was accepted by the 23 wards as well. In line with the negotiation with the union, the TMG and the 23 wards made a new agreement on the devolution in 1994 and issued the Outline of Tokyo-Wards System Reform (the 1994 agreement). This agreement also held IWWD as a significant principle and schemed to devolve the legal responsibility for all of the waste management services including not only collection and transportation, but also the treatment of bulky garbage, waste not for incineration, human waste, and final disposal.

However, it was obvious to all that it was impossible to fulfil this ideal, given the unavailability of landfills in the inland area and unevenly distributed waste disposal facilities across the wards. Consequently, while IWWD was agreed upon in principle, conflict arose over how strictly this principle should be applied in practice, i.e. who should be legally responsible for which waste services and how they were to be performed. As a result of the negotiation, the argument was focused on the self-sufficient incineration again, despite the broad definition of IWWD which covered all waste-related services. Disposal processes other than incineration were excluded from the application of this idea of self-sufficient waste disposal; final disposal was to be entrusted to the TMG, and the bulky garbage, waste not for incineration, and human waste were to be disposed of jointly by local government associations formed by the 23 wards.[27] On the other hand, incineration was agreed to be self-sufficiently performed by each ward.

The focus of this negotiation was on the timing of the devolution. As siting incinerators in every ward was necessary to realise the self-sufficient incineration, the union required a clear roadmap and a guarantee that all the necessary incinerators for IWWD would be constructed. The union insisted that the facilities for IWWD should be built before the legal responsibility was devolved. For the union, imposing tougher conditions was desirable to stop the devolution. On the other hand, the TMG argued that the institutional reform should come first to encourage IWWD. The TMG could not wait until OWOI was achieved, as Governor Suzuki made a public commitment to realise the autonomy reform in 1995 (Miyake, 2006). The TMG and the 23 wards did not like IWWD becoming an obstacle for the autonomy reform[28]; they tried to find a point of compromise to pull off the reform before achieving the OWOI policy.

The 1994 agreement between the TMG and the 23 wards was based on the devolution-first argument. This agreement planned to implement the block incineration system as an interim measure. In this interim system, until enough incinerators for IWWD were ready, the waste from wards without incinerators was to be disposed of in neighbouring wards with spare capacity by making a disposal pact between them. The legal responsibility for incineration would be devolved when enough incinerators for the block incineration were established, that is, when incinerators in Minato, Sumida, Shibuya, and Toshima commenced operation.

However, the union was dissatisfied with this agreement and demanded a tougher condition, which could guarantee the accomplishment of IWWD, for its latent goal was to prevent the devolution. To persuade the union, the TMG had to show its determination to realise IWWD. The schedules for Chuo, Nakano, and Arakawa plants were moved forward as a result of this political negotiation. The union demanded further advancement of the incinerator siting schedule: Chuo plant to start operation by the end of 2000; Chiyoda plant to be under construction by the time of the devolution; the lands to be acquired, agreements with the neighbours to be made, and an environmental impact assessment to be in progress by 1999 in Shinjuku, Nakano, and Arakawa (Tōkyō Seisō Rōdōkumiai & Tōkyō-to, 1994). In the end, the TMG and the 23 wards had to accept this difficult condition to accomplish the autonomy reform. The TMG promised that the devolution would be postponed if this condition were not fulfilled.

Thus, the self-responsibility of each ward became the political focus of the devolution because of the negotiation with the union. Although the self-responsibility of each ward was against its interests, the union persistently called for IWWD in the devolution from its strategic intention to stop the devolution process. Being granted political veto by the MOH, the union was able to make the TMG and the 23 wards accept IWWD as a prerequisite for the devolution. As a result, the self-responsibility of each ward was adopted as a principle of the devolution policies, which also reinforced the idea of siting an incinerator in every ward. The influence of IWWD increased in the early 1990s not only due to the pressure from Koto ward, but also due to the politics of autonomy system reform.

The wards and the autonomy expansion

On the other hand, the 23 wards were more prepared to accept IWWD in both the incinerator siting and the self-responsibility of each ward than they had been in the previous periods, due to the heightened expectations for the autonomy reform and the goal to acquire the legal status as basic local municipalities. They thought that this might be the last chance for them to become basic local municipalities (Tokubetsu-ku Seido Kaikaku Suishin Hombu, 1993). For this political goal, the wards needed to manifest the willingness to accept responsibility for waste management, to show their determination to the MOH. Even before the negotiation with the union began, the 23 wards already unofficially told the TMG in 1990 that they would accept responsibility for incineration if required for the autonomy reform (Tosei Shimpō, 1994b). The autonomy reform gave the wards the motivation to accept IWWD.

In fact, some wards without incinerators showed more positive attitudes towards IWWD even before the 1991 siting plan was announced. In Toshima, the negotiation over the land for the incinerator was already under progress with the cooperation of the ward (Toshima Shimbun, 1991; Toshima-ku-gikai, 1991b).[29] In Nakano, before IWWD was re-adopted by the TMG, the united associations of shopkeepers sent to the government and the council of the ward a petition that requested for an incinerator to be built in front of the Nakano Station (Nakano Eki Shūhen Chiku Seibibu, 1995).[30] This petition stated that having an incinerator was necessary because some wards implied their ownership over incinerators in their wards and another garbage war was about to break out given the increase of waste generation (Nakano-ku-gikai, 1990a). Thus, the ward decided to ask the TMG to build an incinerator in the ward (Nakano-ku-gikai, 1990b). Sumida ward also had been requesting the TMG to site an incinerator in the ward as it was necessary to take a certain degree of responsibility for waste disposal as an independent local autonomy (Sumida-ku-gikai, 1991; Tosei Shimpō, 1991d). In Chuo ward, a local organisation for the redevelopment of the area petitioned to site an incinerator in the ward, as having an incinerator in the ward would make the collection and transportation more efficient if waste management was devolved to the ward (Tsukishima Chiku Saikaihatsu Taisaku Kyōgikai, 1990). The mayor of Chuo also stated in the council that an incinerator should be sited in the ward in light of IWWD (Chūō-ku-gikai, 1991). In the council of Minato ward, there was the argument that an incinerator should be sited for the devolution; the mayor mentioned that the invitation of an incinerator would be considered, as the ward should become as autonomous as a city.[31]

This does not mean that all of the wards were supportive of IWWD. At first, the reaction of the 11 wards without an incinerator was diverse. While some of them showed a positive stance, the mayor of Shinjuku showed a negative attitude towards IWWD. He claimed that the ward had a sewage treatment plant which disposed of sewage from seven wards, and argued that waste treatment and disposal should be performed regionally and that every ward should share different burdens (Shinjuku-ku-gikai, 1991a). Bunkyo and Taito[32] complained of the difficulty to find lands for the incinerators, while they agreed on IWWD and promised to cooperate if a concrete plan would be formulated (Taitō-ku-gikai, 1991; Tosei Shimpō, 1991a). The mayor of Chiyoda stated that the proposed site for an underground plant in the ward was not suitable considering the traffic problems, environmental issues and urban planning design in the central area of Tokyo (Chiyoda-ku-gikai, 1990). Even after the plan was officially announced in 1991, the ward kept a cautious, ambiguous attitude to the project as the proposed underground incinerator was still under financial and technological examination.

However, these wards became rather desperate in searching for an incinerator site, as IWWD was further incorporated into the autonomy expansion reform through the negotiation with the union. The 23 wards published the Third Action Plan[33] in October 1993 and the Basic Way of Waste Management in the 23 wards[34] in April 1994, in order to facilitate the reform. In those documents, the wards stated that they adopted IWWD as a basic principle in waste management and showed their resolution to accept the responsibility as would-be basic local municipalities. On top of that, at the beginning of 1994, the mayors of the 23 wards passed a resolution for the construction of necessary facilities for the devolution (Tokubetsu-ku-chō Kai, 1994). After the 1994 agreement with the TMG was made, the 23 wards issued the Practical Action Plan[35] to show their determination, and the roadmap toward achieving IWWD, in order to persuade the union. This plan included the schedule for incinerators in Shinjuku, Bunkyo and Taito, where no siting plans were included in the 1991 siting plan, and also moved forward the schedules in Nakano and Arakawa.

Thus, the autonomy expansion reform enhanced the 23 wards' willingness to host incinerators. In fact, even Shinjuku, who had been negative at first, became desperate to find a site for an incinerator. The Shinjuku ward council urged the ward government and the TMG to secure a site immediately in response to the negotiation with the union over the devolution as well as pressure from Koto. The council passed a resolution at the end of 1993, promoting the incinerator siting in Shinjuku (Shinjuku-ku Shimbun, 1993). The ward government also started to find possible sites for an incinerator and handed over to the TMG a list of more than 20 candidate sites.[36]

In brief, the heightened expectations for autonomy expansion made the wards more willing to accept the self-responsibility and the incinerator siting which IWWD required. Through the union's intervention, the devolution of incineration turned into a prerequisite for autonomy reform as their long-cherished political goal. This made the wards without an incinerator more eager to host an incinerator. Besides pressure from Koto ward, it was the politics of the autonomy reform that made IWWD influential in waste management policies in this period.

Conclusion

The dominance of IWWD rose once again in the 1990s. IWWD became influential not only in the incinerator siting policies, but also in the institutional reform of waste management. The rising influence of the self-responsibility of each ward in the devolution policies further accelerated incinerator siting. The dominance of IWWD reached its climax when the two engines of this idea were working together.

The pressure from Koto ward, which once had weakened in the previous period, became strong once again as the interest and power of the ward revived as a result of the increased production of waste in the latter half of the 1980s. The new landfill project under the deepening garbage crisis provided the opportunity for Koto to influence waste management policies. The revival of IWWD in this period was championed by the strengthened interest and power of Koto ward.

The ideational legitimacy of OWOI also became more influential than in the second period. Given the increased amount of waste, the TMG needed more incinerators than was planned in the 1980s to achieve AWI. This urgent necessity for new incinerators was further heightened through the quick depletion of the existing landfill and the policy goal of not bringing un-treated waste into the new landfill. IWWD became cognitively appealing to the TMG once again as a policy solution to the problem.

The economic situation, as an exogenous environment, played a central role in strengthening the idea of siting incinerators in every ward. Some of the changes in the other variables mentioned above were caused by the bubble economy. Koto's interest in IWWD revived with the increased waste generation through the economic boom, while the resultant quick depletion of landfills and the urgent need of a new landfill empowered the ward in the policy-making process. IWWD was recognised by the TMG as a policy solution in part due to the increased amount of garbage resulting from the economic boom. The huge budgetary funds accumulated during the period of economic prosperity made the incinerator projects in the 1991 affordable. The economic boom in the late 1980s created a favourable environment for IWWD once again.

It is true that the prominence of IWWD could have been undermined by the bubble burst in the 1990s and the rise of Sustainable Waste Management as the new policy paradigm. However, these exogenous changes had not yet fully impacted IWWD, given the estimation that the amount of waste would keep increasing in the future, with the rather optimistic prospect of an economic recovery. In other words, there was a time gap between the changes in the exogenous environments and the decline of the idea in the policies. It was not until 1997 that the impact of the economic downturn and the advancement of SWM significantly affected the dominance of IWWD.

The interest and power of the Tokyo Cleaning Workers' Union engendered the political engine for the rise of the idea that each ward should be institutionally responsible for disposing of its waste. In the 1970s and the 1980s, the union's interest and power had prevented IWWD from influencing the devolution. In the 1990s, however, the union persistently called for IWWD in the devolution from its strategic intention to stop this institutional change; the union tried to frustrate the devolution by adopting IWWD as a tough condition on this institutional reform, as siting incinerators in every ward was expected to be very difficult to accomplish. The union was able to make the TMG and the 23 wards accept IWWD as a prerequisite for this institutional reform, as it was granted by the MOH a political veto power over the devolution and the entire autonomy reform of the 23 wards.

In a sense, the union took advantage of the ambivalence in the ideational legitimacy of IWWD; it was recognised as normatively compelling but cognitively not convincing. The self-responsibility of each ward was normatively compelling as the autonomy expansion movement reached its culmination in this period. The more autonomous and independent the 23 wards were trying to become, the more normatively legitimate the argument became that a ward should take care of its own waste. This heightened normative resonance with the autonomy of the 23 wards made IWWD undeniable for the 23 wards as well as the TMG. On the other hand, devolving waste disposal to each ward was not convincing to all parties; it was recognised as problem-causing rather than problem-solving. The union called for IWWD knowing that it was cognitively problematic but normatively irrefutable. The strategy of the union to prevent the devolution was facilitated by this ambivalence in ideational legitimacy.

Taken together, the political drive for the institutional responsibility of each ward was produced through the interaction between the ambivalence in the ideational legitimacy, the conflicting interests among the key actors, the power of the union, and the irresolute attitude of the MOH to the autonomy reform as an exogenous political environment. The MOH required the devolution of waste management and the agreement of the concerned parties as conditions for autonomy reform, which had been the political goal for the 23 wards and the TMG. The union's interest in frustrating the devolution, interplaying with the ambivalence

in ideational legitimacy and the veto granted in the political process of the autonomy reform, resulted in the strategy of preventing the devolution by calling for
IWWD. This strategy worked not only because the 23 wards and the TMG had to
persuade the union to achieve their political goal of devolution, but also because
IWWD was normatively irrefutable as they advocated the autonomy of the wards.
The idea of the self-responsibility of each ward became influential in the devolution policies through this interaction of the variables, although all of the parties
doubted its policy rationale.

Thus, the dominance of IWWD came to a climax in this period. The influence
of OWOI revived through the interaction between the variables centred on the
economic conditions. The idea of the self-responsibility of each ward also became
influential through the complicated interaction of the four variables and further
accelerated the incinerator siting policy. Yet, this dominance did not last for long.
The next chapter elucidates the decline of IWWD in the latter half of the 1990s
and its abandonment in the 2000s.

Notes

1 Seisō Kōjō Kensetsu Keikaku [清掃工場建設計画] (Tōkyō-to Seisō Kyoku, 1991c).
2 Chiyoda, Chuo, Minato, Shinjuku, Bunkyo, Taito, Sumida, Shibuya, Nakano, Toshima
   and Arakawa.
3 The TMG expected these two projects to be realised soon.
4 Toku seido Kaikaku ni Kansuru Matome [都区制度改革に関するまとめ] (Toku
   Kyōgikai & Toku Seido Kaikaku Suishin Iinkai, 1994).
5 On the other hand, the Phoenix Project, a cross-jurisdictional landfill project in
   Tokyo Bay jointly utilised by prefectures and cities, had made no progress and was
   not expected to be completed on schedule. The Phoenix Project, based on the Act on
   Bay Area Marine and Environment Consolidation Centres (Kōiki Rinkai Kankyō Sentā
   Hō [広域臨海環境整備センター法]) in 1981, aimed at constructing huge landfills
   in Tokyo Bay and Osaka Bay to deal with waste from the metropolitan areas. While
   the project was materialised in the Osaka Bay area, the one in Tokyo Bay failed due to
   disagreement among the prefectures and cities involved.
6 Tōkyō-to Kōwan Shingikai [東京都港湾審議会].
7 The other two were at off-shore sites in Kasai and Haneda.
8 In the first garbage war, Koto required new landfills at off-shore sites in Haneda and
   Kasai to disperse the burden of landfills.
9 Experiencing serious pollution through the 1950s and the 1960s, environmental
   assessment had been conducted by ministries and governmental offices for large-
   scale projects since the early 1970s. After an attempt to legislate environmental
   assessment failed in 1983, the Environmental Impact Assessment Guideline was
   decided by the cabinet of the government in 1984. At a national level, environmental
   assessment had been implemented according to this guideline until the enactment
   of the Environment Impact Assessment Act in 1997. On the other hand, some prefectures and cities had institutionalised environmental assessment by making local
   ordinances or guidelines since the 1970s. In Tokyo, the Tokyo Environmental Impact
   Assessment Ordinance passed the Tokyo Metropolitan Assembly in 1980 after the
   failed attempt by Governor Minobe at the end of the 1970s. The new landfill was
   assessed according to the national guideline and the local ordinance (Setsuritsu
   Nijūgo Shūnen Kinen Jigyō Zikkō Iinkai Kankyō Asesumento Shi Shippitsu Shō
   Iinkai, 2003).

10 The Act on Reclamation of Publicly-owned Water Surface was amended in 1973 and the regulation on reclaiming public water was tightened. The amendment required hearing the opinion of the mayor of concerned local municipalities and the resolution of the local council in the process of issuing the license.

11 For example, see Shinjuku-ku-gikai (1991b, 1991c).

12 Furthermore, the ward sent an open letter to the members of the Harbour Advisory Committee and complained of the burden that Koto had suffered and denounced its myopic view on waste disposal and IWWD (Kōtō-ku-gikai, 1991c).

13 The others were the Long-Term Vision of Waste Management (Haikibutsutō no Shori Shobun no Chōkiteki Tembō ni tsuite [廃棄物の処理処分の長期の展望について]) (Tōkyō-to Seisō Kyoku, 1991b) and Waste Reduction Action Plan (Gomi Genryōka Kōdō Keikaku [ごみ減量化行動計画]) (Tōkyō-to Seisō Kyoku, 1991a).

14 It was customary to have a meeting between the TMG and the ward (Toku Kyōgikai [都区協議会]) and to make an agreement with the ward before holding the Harbour Advisory Committee (Tōkyō-to-gikai, 1991a).

15 The other requests were the following: reduction of the number of garbage trucks passing through the ward; advancement of reduction and recycling; facilitation of the construction of intermediate facilities for waste not for incineration; and measures to reduce the surplus soil being sent to the landfill.

16 The governor back then was Yukio Aoshima, who won the election in 1995 after Suzuki stepped down.

17 Shinjuku had found a site for an incinerator by that time.

18 The landfill consisted of seven blocks from A to G, which would be reclaimed in turn (Kōtō-ku-gikai, 1996; Tosei Shimpō, 1995).

19 Another reason why the ward had to agree on the siting project was that starting operation of the new landfill was a condition for the devolution of waste management imposed through the negotiation with the union (Kōtō-ku-gikai, 1998).

20 Haikibutsutō no Shori Shobun no Chōkiteki Tembō ni tsuite [廃棄物等の処理処分の長期的展望について] (Tōkyō-to Seisō Kyoku, 1991b).

21 Gross prefectural expenditure at constant prices based on Prefectural Accounts 1990–2003 (93SNA, benchmark year=1995).

22 The Waste Management and Public Cleaning Act (Haikibutsu no Shori oyobi Seiō ni Kansuru Hōritsu [廃棄物の処理及び清掃に関する法律]) was amended in 1991 and 1997. Previously, the purpose of this act had been to properly dispose of waste without causing pollution. However, these amendments emphasised the promotion of waste reduction and recycling. The Act on the Promotion of Effective Utilization of Resources (Shigen no Yūkō na Riyō no Sokushin ni Kansuru Hōritsu [資源の有効な利用の促進に関する法律]) was issued in 1991. Responding to the amendment in 1991 and the enactment of the act, the TMG established a new ordinance on waste disposal and utilisation of recyclable resources (Tōkyō-to Haikibutsu no Shori oyobi Sairiyō ni Kansuru Jōrei [東京都廃棄物の処理及び再利用に関する条例]). This ordinance introduced the idea of restraining waste generation in production, distribution and consumption as a basic principle. The plans in waste management in this period, such as the Third Tokyo Long-Term Plan in 1990, the Long-Term Vision of Waste Management in 1991, and the Waste Reduction Action Plan in 1991, were grounded on the same idea.

23 The Ministry of Home Affairs was promoted from the Office of Home Affairs in 1960, but merged into the Ministry of Internal Affairs and Communications in the central government reform in 2001.

24 Dai Nijūniji Chihō Seido Chōsakai [第二十二次地方制度調査].

25 The other party was the private businesses which were commissioned waste transportation services.

26 Seisō Jigyō no Gutaiteki Arikata ni tsuite [清掃事業の具体的あり方について] (Tōkyō-to, 1994).

27 The 1994 agreement stated that the construction of intermediate treatment facilities for bulky waste and waste not for incineration in every ward would be considered in the future; they were to be disposed of regionally for the time being.
28 Even the mayor of Koto ward was apprehensive that imposing IWWD as a condition for the devolution would make the autonomy reform difficult to attain (Murohashi, 1998).
29 Toshima ward even insisted that the capacity of the proposed incinerator was too small to dispose of all the waste for incineration in the ward and asked the TMG to expand the capacity from 300 tonnes/day to 400 tonnes/day (Toshima-ku-gikai, 1991a, 1992).
30 The Police Academy in front of the Nakano Station was to be relocated and an incinerator construction was proposed as a part of the redevelopment project there.
31 For instance, see a statement of Takiko Otaki (Minato-ku-gikai, 1990).
32 Neighbourhood associations in Taito also proposed to invite an incinerator in 1992 (Tōkyō-to-gikai, 1994a, 1994b; Ueno Chiku Chōkai Rengō Kai, 1992).
33 Dai Sanji Kōdō Keikaku [第三次行動計画] (Tokubetsu-ku Seido Kaikaku Suishin Hombu, 1993).
34 Tokubetsu-ku ni okeru Seisō Jigyō no Kihonteki Arikata [特別区における清掃事業の基本的あり方] (Tokubetsu-ku Seido Kaikaku Suishin Iinkai, 1994).
35 Gutaiteki Kōdō Keikaku [具体的行動計画]. This plan introduced a roadmap to achieve IWWD in which OWOI was to be achieved by 2011, with completion in Chiyoda in 2002, Shinjuku, Nakano and Arakawa in 2006, and Taito and Bunkyo in 2011 (Toku Seido Kaikaku Suishin Iinkai, 1994).
36 The site was selected in 1994.

References

Asahi Shimbun. (1991a). Kōtō-ku ga To e Kōgi Kōdō. *Asahi Shimbun Sha*, 1.8.1991.
Asahi Shimbun. (1991b). Oshitsuke wa Gomen: Sainen? Gomi Sensō. *Asahi Shimbun Sha*, 1.8.1991.
Asahi Shimbun. (1991c). Toshin no Chika Shorijō ni Sampi. *Asahi Shimbun Sha*, 21.11.1991.
Chiyoda-ku-gikai. (1990). Chiyoda-ku-gikai Dai Nikai Teireikai: Heisei Ninen Rokugatsu Nanoka.
Chūō-ku-gikai. (1991). Chūō-ku-gikai Kaigiroku: Heisei Sannen Dai Sankai Teireikai.
Dai Nijūniji Chihō Seido Chōsakai. (1990). *Dai Nijūniji Chihō Seido Chōsakai Tōshin.*
Jinno, N. (1995). Toshi Keiei no Hatan kara Saiken e. In N. Jinno (ed.), *Toshi o Keiei Suru* (pp. 67–138). Tokyo: Toshi Shuppan.
Kōtō-ku-gikai. (1989a). Gomi no Shin Shobunjō ni Kansuru Ikensho.
Kōtō-ku-gikai. (1989b). Seisō Mondai Toku Kyōgikai ni okeru Iinkai Iken.
Kōtō-ku-gikai. (1989c). Toku Kyōgikai ni okeru Iinkai Iken.
Kōtō-ku-gikai. (1991a). Kōtō-ku Chisaki Shin Kaimen Shobunjō Kensetsu Keikaku ni Hantai suru Yōsei.
Kōtō-ku-gikai. (1991b). Kōtō-ku-gikai Kaigiroku Heisei Sannen Dai Yonkai Teireikai.
Kōtō-ku-gikai. (1991c). Shin Kaimen Shobunjō Hantai ni Kansuru Kōkai Shitsumonjō.
Kōtō-ku-gikai. (1991d). Shin Kaimen Shobunjō Mondai ni Kanrensuru Seisō Kōjō Kensetsu Keikaku ni taisuru Iken.
Kōtō-ku-gikai. (1991e). Toku Kyōgikai ni okeru Seisō Kōwan Tokubetsu Iinkai no Torimatome (Jūichigatsu Nijūkunichi).
Kōtō-ku-gikai. (1992a). Shin Kaimen Shobunjō no Keika ni tsuite.

Kōtō-ku-gikai. (1992b). Tōkyōwan Dai Goji Kaitei Kōwan Keikaku no Ichibu Henkō An ni tsuite Seisō Kōwan Tokubetsu Iinkai no Matome.

Kōtō-ku-gikai. (1996). Shin Kaimen Shobunjō Mondai no Keika ni tsuite.

Kōtō-ku-gikai. (1998). Kusei Taisaku, Seisō Kōwan Rinkaibu Taisaku Tokubetsu Iinkai no Matome.

Minato-ku-gikai. (1990). Minato-ku-gikai Kaigiroku: Heisei Ninen Dai Sankai Teireikai.

Miyake, H. (2006). Nisennen Toku Seido Kaikaku Seisō Jigyō Ikan no Seiritsu to sono Kyōgi Katei. *TIMR Research Paper*, 1, 1–36.

Murohashi, A. (1998). *Waga Machi Kōtō ni Ai to Hokori o*. Tokyo: Sampō Sha Insatsu.

Nakano Eki Shūhen Chiku Seibibu. (1995). Keisatsu Daigakkōtō Iten Shikichi no Seisō Kōjō Kensetsu no Kentō Keika.

Nakano-ku-gikai. (1990a). Chinjō Bunsho Hyō: Nakano-ku-gikai Kaigiroku Heisei Ninen Dai Ikkai Teireikai.

Nakano-ku-gikai. (1990b). Nakano-ku-gikai Kaigiroku: Heisei Ninen Dai Sankai Teireikai.

Narita, Y. (1998). Toku Seido Kaikaku ni Kansuru Chihōjichi Hō Kaisei ni tsuite. *Chihō Jichi*, 607, 2–10.

Setsuritsu Nijūgo Shūnen Kinen Jigyō Zikkō Iinkai Kankyō Asesumento Shi Shippitsu Shō Iinkai. (2003). *Nihon no Kankyō Asesumento Shi*. Tokyo: Nihon Kankyō Asesumento Kyōkai.

Shinjuku-ku Shimbun. (1993). Seisō Kōjō o Kunai ni. *Shinjuku-Ku Shimbun Sha*, 15.12.1993.

Shinjuku-ku-gikai. (1991a). Shinjuku-ku-gikai Kessan Tokubetsu Iinkai: Heisei Sannen Jūichigatsu Jūyokka.

Shinjuku-ku-gikai. (1991b). Shinjuku-ku-gikai Kessan Tokubetsu Iinkai: Heisei Sannen Jūichigatsu Nijūgonichi.

Shinjuku-ku-gikai. (1991c). Shinjuku-ku-gikai Kessan Tokubetsu Iinkai: Heisei Sannen Jūichigatsu Nijūichinichi.

Sumida-ku-gikai. (1991). Sumida-ku-gikai Kaigiroku: Heisei Sannen Dai Ikkai Teireikai.

Taitō-ku-gikai. (1991). Taitō-ku-gikai Kaigiroku Heisei Sannen Dai Sankai Teireikai.

Toku Kyōgikai & Toku Seido Kaikaku Suishin Iinkai. (1994). *Toku Seido Kaikaku ni Kansuru Matome*.

Toku Seido Kaikaku Suishin Iinkai. (1994). *Gutaiteki Kōdō Keikaku no Gaiyō*.

Toku Seido Kentō Iinkai. (1986). *Toku Seido Kaikaku no Kihon Hōkō*.

Tokubetsu-ku Seido Kaikaku Suishin Hombu. (1993). *Dai Sanji Kōdō Keikaku*.

Tokubetsu-ku Seido Kaikaku Suishin Iinkai. (1994). *Tokubetsu-ku ni okeru Seisō Jigyō no Kihonteki Arikata*.

Tokubetsu-ku-chō Kai. (1994). Ketsugi Sho.

Tōkyō Nijūsan-ku Seisō Ichibu Jimu Kumiai. (2006). *Ippan Haikibutsu Shori Kihon Keikaku*.

Tōkyō Seisō Rōdōkumiai. (1999). *Tōkyō Seisō Rōdōkumiai Gojūnen-shi*.

Tōkyō Seisō Rōdōkumiai & Tōkyō-to. (1994). Kankeisha toshiteno Tōkyō Seisō Rōdō Kumiai to Totōkyoku no Kōshō Gijiroku no Kakunin oyobi Oboegaki.

Tōkyō Tokusyokuin Rōdōkumiai. (1993a). Dai Jūnikai Seiō Mondai ni Kansuru Semmon Iinkai.

Tōkyō Tokusyokuin Rōdōkumiai. (1993b). Dai Jūsankai Seiō Mondai ni Kansuru Semmon Iinkai.

Tōkyō Tokusyokuin Rōdōkumiai. (1993c). Dai Sankai Seiō Mondai ni Kansuru Semmon Iinkai.

Tōkyō Tokusyokuin Rōdōkumiai. (1993d). Dai Yonkai Seiō Mondai ni Kansuru Semmon Iinkai.

Tōkyō Tokusyokuin Rōdōkumiai. (1995). *Toku Seido Kaikaku o Meguru Tatakai no Ayumi.*

Tōkyō-to. (1982). *Tōkyō-to Chōki Keikaku: Mai Taun Tōkyō 21 Seiki o Mezashite.*

Tōkyō-to. (1993). *Seisō Jigyō no Arikata ni tsuite.*

Tōkyō-to. (1994). *Seisō Jigyō no Gutaiteki Arikata ni tsuite.*

Tōkyō-to Gyōzaisei Kaikaku Suishin Hombu. (1996). *Tōkyō-to Zaisei Kenzenka Keikaku.*

Tōkyō-to Kikaku Hōdō Shitsu. (1981). *Mai Taun Tōkyō 81.*

Tōkyō-to Kikaku Shingi Shitsu. (1987). *Mai Taun Tōkyō 87.*

Tōkyō-to Kikaku Shingi Shitsu. (1990). *Dai Sanji Tōkyō-to Chōki Keikaku.*

Tōkyō-to Kōwan Shingikai Kaimen Shobunjō Kentō Bukai. (1991a). *Haikibutsutō no Aratana Kaimen Shobunjō Seibi no Kihon Hōshin ni tsuite Chūkan Hōkoku.*

Tōkyō-to Kōwan Shingikai Kaimen Shobunjō Kentō Bukai. (1991b). *Haikibutsutō no Aratana Kaimen Shobunjō Seibi no Kihon Hōshin ni tsuite Saishū Hōkoku.*

Tōkyō-to Seisō Kyoku. (1991a). *Gomi Genryōka Kōdō Keikaku.*

Tōkyō-to Seisō Kyoku. (1991b). *Haikibutsutō no Shori Shobun no Chōkiteki Tembō ni tsuite.*

Tōkyō-to Seisō Kyoku. (1991c). *Seisō Kōjō Kensetsu Keikaku.*

Tōkyō-to Seisō Kyoku. (2000). *Tōkyō-to Seisō Jigyō Hyakunen-shi.* Tokyo: Tōkyō-to Kankyō Seibi Kōsha.

Tōkyō-to Seisō Shingikai. (1990a). *Seisō Jigyō no Kongo no Arikata ni tsuite: Chūkan Tōshin.*

Tōkyō-to Seisō Shingikai. (1990b). *Seisō Jigyō no Kongo no Arikata ni tsuite: Saishū Tōshin.*

Tōkyō-to-gikai. (1991a). Tōkyō-to-gikai Jūtaku Kōwan Iinkai: Heisei Sannen Jūnigatsu Jūsannichi.

Tōkyō-to-gikai. (1991b). Tōkyō-to-gikai Kensetsu Seisō Iinkai: Heisei Sannen Jūichigatsu Jūyokka.

Tōkyō-to-gikai. (1994a). Tōkyō-to-gikai Kensetsu Seisō Iinkai: Heisei Rokunen Sangatsu Jūgonichi.

Tōkyō-to-gikai. (1994b). Tōkyō-to-gikai Yosan Tokubetsu Iinkai: Heisei Rokunen Sangatsu Jūgonichi.

Tosei Shimpō. (1991a). Kankei Jūniku no Taiō. *Tosei Shimpō Sha*, 1.11.1991.

Tosei Shimpō. (1991b). Kōtō-ku-gikai Dainiji Gomi Sensō Sengen?. *Tosei Shimpō Sha*, 26.7.1991.

Tosei Shimpō. (1991c). Kyūjūichinen To Sōgō Jisshi Keikaku: Tokuchō to Kongo no Kadai 3. *Tosei Shimpō Sha*, 13.12.1991.

Tosei Shimpō. (1991d). Towareru Tokubetsu-ku 2. *Tosei Shimpō Sha*, 16.11.1991.

Tosei Shimpō. (1994a). Shinjuku-ku Ichigaya Hommura ni Seisō Kōjō Kensetsu e. *Tosei Shimpō Sha*, 11.10.1994.

Tosei Shimpō. (1994b). Toku Seido Kaikaku no Yukue 8. *Tosei Shimpō Sha*, 26.4.1994.

Tosei Shimpō. (1995). Kujū no Omoi de Kibishii Ketsudan. *Tosei Shimpō Sha*, 10.11.1995.

Toshima Shimbun. (1991). Fukutoshin no Ittōchi ni Seisō Kōjō. *Toshima Shimbun Sha*, 24.9.1991.

Toshima-ku-gikai. (1991a). Toshima-ku-gikai Kaigiroku: Heisei Sannen Jūichigatsu Nijūichinichi.

Toshima-ku-gikai. (1991b). Toshima-ku-gikai Kaigiroku: Heisei Sannen Kugatsu Sanjūnichi.

Toshima-ku-gikai. (1992). Toshima-ku-gikai Kaigiroku: Heisei Yonen Kugatsu Nijūkunichi.

Tsukishima Chiku Saikaihatsu Taisaku Kyōgikai. (1990). Chūō-ku Nai Asashio Unga eno Seisō Kōjō Kensetsutō ni Kansuru Seigan.

Ueno Chiku Chōkai Rengō Kai. (1992). Taitō-ku Nai no Kokuyūchi (Gen Shihō Shūshūjo) ni Chikashiki Seisō Kōjō oyobi Kōen o Kensetsu Surukoto ni Kansuru Seigan.

6 Decline and abandonment of IWWD

The dominance of In-Ward Waste Disposal (IWWD) started declining in 1997 after it reached what would be its culmination in the early half of the 1990s. The goal for incinerator siting, which was planned in 1991 and accelerated through the previous period, was revised downwards. Furthermore, in 1998, the shift to the self-sufficient incineration system was rescheduled. The influence of the idea of distributive justice further declined in the 21st century. In 2003, the 23 wards, who were in charge of waste management since the devolution reform in 2000, concluded that there was no need for new incinerators. At the same time, they also abandoned the shift to the self-sufficient incineration system in which each ward was to perform incineration of its own waste.

This chapter explains this decline and abandonment of IWWD in the fourth period. The decreasing amount of waste undermined the cognitive legitimacy of IWWD as a policy solution, and also alleviated the pressure from Koto ward. The financial crisis damaged the feasibility of the One Ward One Incinerator (OWOI) policy. The long economic recession, the advancement of the 3Rs under Sustainable Waste Management (SWM) and the dioxin crisis created a disadvantageous environment for the idea of siting incinerators in every ward. This decline of the idea of siting incinerators in every ward led to the decline in the policy of the self-responsibility of each ward.

One Ward One Incinerator in decline

Revision of the 1991 siting plan in 1997

The dominance of IWWD in siting policies started declining in 1997. The TMG revised the incinerator siting plan and started retreating from IWWD, postponing the achievement of the OWOI policy to a future date. The 1991 incinerator siting plan projected that 20 wards would have at least one incinerator by 2011, although the plan did not propose any practical siting projects for the other three wards. This schedule was accelerated through the pressure from Koto and the negotiation with the union over the devolution reform. The site in Shinjuku, which had been not specified in the 1991 plan, was finally selected in 1994. The schedule for Nakano and Arakawa was moved forward to ensure completion by 2006. The TMG also promised to the union that the incinerators siting would be implemented on schedule, stating that the Chiyoda project was to start construction, and the environmental assessments would be underway in Shinjuku, Nakano, and Arakawa by the time of the devolution of waste management in 2000. Until 1996, the TMG maintained this schedule, stating in the Assembly that negotiations were underway in Shinjuku, Nakano, and Arakawa to start construction by 2002 (Tōkyō-to-gikai, 1996), while the project in Chiyoda, the underground incinerator, was being reconsidered.

The turning point came in 1997 when the TMG issued two policy drafts to revise the 1991 siting plan (Tōkyō-to Seisō Kyoku, 1997a, 1997b). These drafts reconsidered the projects after the Chuo plant as shown in Table 6.1. The lands for Shinjuku, Nakano, and Arakawa would be used for recycling-related facilities for the time being and the incinerators would be rescheduled to a time that would be dependent on the prevailing trend of waste growth.

These revisions were incorporated into the Tokyo Slim Plan 21,[1] released at the end of 1997. In this plan, the completion of the OWOI policy was postponed further. Even though the Shibuya and Chuo plants were to be completed, the revision virtually shelved the projects for the remaining six wards (Figure 6.1). The Tokyo Slim Plan 21 still referred to IWWD as its basic principle and stated that OWOI would be achieved by around 2018 if socio-economic conditions remained unchanged, with projects being completed in Shinjuku, Nakano and Arakawa

Table 6.1 Schedule of incinerator projects in the 1997 revision

Projects	1991 siting plan (1994 advancement)		1997 revision	
	Construction	Operation	Construction	Operation
Sumida	1993	1996	started in 1994	1997
Setagaya	1993	1996	cancelled	
Incinerator vessel	1993	1996	cancelled	
Minato	1994	1998	started in 1995	1998
Toshima	1994	1998	started in 1995	1999
Shibuya	1994	1998	1998	2001
Chuo	1997	2001 (2000)	1998	2001
Chiyoda	1997	2002	2015 (site not specified)	2018
Nakano	2004 (2002)	2008 (2006)	2010	2013
Arakawa	2007 (2002)	2011 (2006)	2010	2013
Shinjuku	not specified (2002)	not specified (2006)	2010	2013
Taito	not specified (searching for a site)		2012 (site not specified)	2015
Bunkyo	not specified (searching for a site)		2012 (site not specified)	2015

Source: Toku Kyōgikai & Toku Seido Kaikaku Suishin Iinkai (1994), Toku Seido Kaikaku Suishin Iinkai (1994), Tōkyō Seisō Rōdōkumiai & Tōkyō-to (1994), and Tōkyō-to Seisō Kyoku (1991, 1997a, 1997b 1997c).

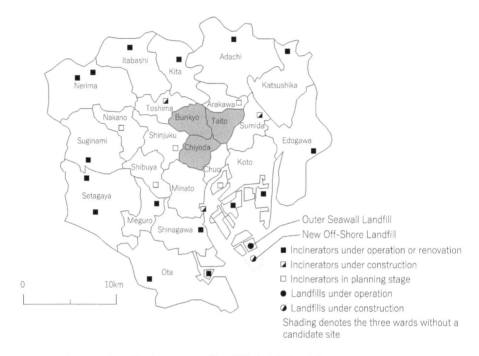

Figure 6.1   Location of incinerators and landfills in 1997 revision.

Adapted from a map of Tokyo by CraftMap (URL: http://www.craftmap.box-i.net/) based on Tōkyō-to Seisō Kyoku (1997c).

in 2013, Bunkyo and Taito in 2015, and Chiyoda in 2018. However, compared to the 1991 siting plan, the progress of incinerator siting had obviously slowed down.

What weakened the influence of IWWD?

The siting plan was revised neither because of limited land availability nor local opposition. In the 1970s, the limited land availability and persistent local opposition made the implementation difficult and led to the revision in the Tokyo Mid-Term Plan 1974. On the other hand, the declining influence of OWOI in the latter half of the 1990s cannot be well explained by land unavailability or local opposition.

This does not mean sites for incinerators were easier to find in this period. While the 1991 siting plan announced the specific candidate sites for eight projects out of the 11 wards without an incinerator, the other three wards, i.e. Shinjuku, Bunkyo, and Taito, were left without any candidate sites. While Shinjuku somehow managed to find a site in 1994, lands were not secured in Bunkyo and Taito, despite efforts by the TMG and the pressure from Koto and the union. In Chiyoda, the underground project was proposed because of the difficulty to find a site above ground; no new candidate site was suggested after the reconsideration of the underground project in 1994 due to technological and cost problems. As a result, three wards, i.e. Chiyoda, Bunkyo and Taito, were left without any candidate sites. Thus, the difficulty in securing sites was still a constraint on realising OWOI in the 1990s.

However, the three projects in Shinjuku, Nakano and Arakawa were postponed, and cancelled later in 2003, even though the land acquisition for them was confirmed after gaining consent from the landowners. In Shinjuku, it was agreed in 1994 that a part of national land would be purchased for the incinerator project. The candidate site in Nakano was also a part of the National Police Academy, which was earmarked to be relocated. Although there had been issues on the specifications of the plant, the land acquisition was scheduled. In Arakawa, the negotiation over the land was proceeding with the cooperative attitude of a private company which owned the land. The 1997 revision put off the incinerator projects there, and the lands were to be used as garbage stockyards for the time being. Thus, the difficulty in finding lands is not a sufficient explanation for the slowdown of the incinerator siting projects.

Local opposition did not significantly disturb the progress of the OWOI policy either. This was in contrast to the first garbage war in which local resistance to the projects prevented the progress of IWWD. This does not mean that there was no local opposition to the incinerator projects in the 1990s. Local opposition movements against the incinerators arose at most of the proposed sites.[2] As will be detailed below, there were protests against the incinerators in Chuo, Minato, Shinjuku, Shibuya, and Toshima wards. Even Nakano, where the ward and the neighbours' association invited the siting of an incinerator, met local opposition. However, the delay of the projects caused by these local protests was not sufficient to affect the strength of IWWD.

Reasons to protest against the incinerators

The siting projects were protested in part because they were against the interests of the locals. Traffic problems were a main concern as hundreds of garbage trucks would drive through a community once an incinerator was sited. The fear of environmental pollution from a plant was also prevalent among the local opposition. Negative impact on community development was another common reason, especially in Chuo, Shinjuku, Shibuya and Toshima, where those who had interests in the development of the areas, such as shopkeepers and local businesses, opposed the projects.

The projects were challenged from the viewpoint of procedural fairness as well. The lack of sufficient consultation with the neighbours was often mentioned in the conflicts. The locals around the project in Shinjuku accused the TMG of not having enough consultation with them. In Chuo and Toshima, the siting process was denounced as undemocratic, and accused of ignoring the local people. In Shibuya, the site selecting process was criticised, for the proposed site was different from the one in Yoyogi Park which was initially chosen voluntarily by the citizens committee during the first garbage war. The local opposition claimed that the site selection process for Yoyogi Park was suitable and consensual, and denounced the site proposed by the TMG.

Some projects were opposed in light of IWWD. The interim report from the Tokyo Waste Management Advisory Committee in 1990 recommended utilising lands owned by the TMG, such as the sites of the Tokyo Metropolitan University in Meguro and Setagaya wards (Tōkyō-to Seisō Shingikai, 1990). However, right after the report was published, a protest arose in Meguro where one incinerator, which was agreed upon after intense opposition in the 1980s, was already under construction. The ward council passed the resolution against siting another incinerator in the ward, contending that no more incinerators were acceptable from the perspective of IWWD (Meguro-ku-gikai, 1990; Tosei Shimpō, 1990, 1991). Facing the opposition, the TMG decided to choose the other site of the university in Setagaya ward, and this project was included in the 1991 siting plan (Setagaya-ku-gikai, 1991).

In Setagaya, however, the project was accused of being inconsistent with IWWD, for there were already two incinerators in the ward. The opposition movements argued that the amount of garbage in Setagaya was only two-thirds of the capacity of the two incinerators already existing in the ward and hence there was no need for a third one, in light of IWWD. The opposition collected more than 24,000 signatures and petitioned against the project (Fukuwatari, 1992; Ichishima, 1992; Mita, 1992). Facing this rise of protest, the mayor of the ward also declared his opposition to the project in November 1992 (Setagaya-ku-gikai, 1992).

Another example was the incinerator vessel at the site of the existing landfill next to Koto ward. As noted in the previous chapter, the ward intensely opposed this project in the campaign against the new landfill siting. In Koto ward, there were already two incinerators in operation, which treated waste from the surrounding wards. For Koto, accepting another burden was out of the question, in light of IWWD. In fact, in order to achieve All Waste Incineration (AWI) as soon as possible, these projects were planned at the sites where the TMG believed the projects would proceed quickly without considering IWWD.

IWWD worked negatively on the incinerator siting in Shibuya as well. The TMG proposed an incinerator of only 200 tonnes/day capacity because of the limited space at the proposed site, while the amount of waste in Shibuya was around 500 tonnes/day.[3] Some of those opposing the project, therefore, argued that the proposed site was inappropriate and not large enough to accomplish IWWD in Shibuya. This issue of the capacity was linked to the problem of the site selection process noted above. While the site proposed by the TMG was too small for IWWD, argued the local opposition, the one in Yoyogi Park was large enough to construct an incinerator which could deal with all waste for incineration in the ward. In these cases, the idea of distributive justice provided the legitimacy to the resistance movements rather than to the projects.

Anti-incineration movements

Furthermore, the emergence of anti-incinerationism, an idea which denied the very necessity of incinerators, challenged the idea of siting incinerators in every ward. Anti-incinerationism attacked incinerationism, which had been the paradigmatic idea in waste management, by challenging its technological belief in incineration. Incineration had initially been regarded as the best way to dispose of waste, due to sanitation concerns and later due to the shortage of landfill space. In contrast, anti-incinerationism was based on the technological disbelief in and strong aversion to incineration.

Anti-incinerationism criticised incinerationism for its end-of-pipe centred waste management. Sustainable Waste Management (SWM), which was becoming increasingly influential on waste management policies, provided the theoretical backbone for anti-incineration. SWM provided an alternative resolution to the garbage crisis apart from expanding incineration. Instead of enlarging the end-of-pipe disposal capacity, anti-incineration emphasised source reduction. Anti-incineration criticised the expansion of incineration capacity for being against a recycling-oriented society, arguing that having large disposal capacity would discourage reduction and recycling efforts. In the first garbage war, having enough incineration capacity was believed to be the only solution to the garbage crisis. Even the local opposition admitted the need of incinerators, though they did not welcome the projects in their communities. In contrast, anti-incinerationism, inspired by SWM and source reduction, cast doubt on the very necessity of incinerators.

Furthermore, anti-incinerationism was based on a strong aversion to incineration as a source of pollution. Some advocates of this idea even denounced incinerators as chemical substance producing factories.[4] In particular, the dioxin crisis discredited incineration in the 1990s. Dioxin pollution from incinerators was attracting much public attention in this period. Since the detection of dioxin from the ashes of municipal incinerators was publicised in 1983, the fear of the toxic substance was spreading among neighbours living around existing and proposed incinerator sites. Dioxin pollution was added to the vocabulary of the local opposition and caused the sentiment against incinerators by associating them with the toxic substance sprayed during Vietnam War – described as the worst poisoning in human history. For those against incinerationism, the very existence of incinerators was no longer tolerable.

From the perspective of anti-incinerationism, IWWD was an idea which encouraged the construction of unnecessary, pollutive incinerators. Anti-incinerationism movements condemned IWWD for being tightly connected with constructing more incinerators.[5] In the early 1970s, IWWD was lionised as the shift from dumping-oriented disposal to incineration-oriented disposal was a pressing issue. For anti-incinerationism, IWWD was an out-of-date idea which believed that expanding the end-of-pipe disposal capacity could solve waste problems.[6] Thus, the rise of anti-incinerationism among the neighbours, backed by SWM and the idea of source reduction, undermined the ideational legitimacy of IWWD which was tightly connected to the construction of a large number of incinerators. Objections based on anti-incinerationism were prevalent in the most of local opposition during this period; it was seen in Minato, Shinjuku, Shibuya, Setagaya, Nakano, and Toshima wards.

Local opposition relatively weak

However, these local protests did not significantly delay the siting of incinerators. Institutionally, there were opportunities for citizens to express their opinions in the process of urban planning decision and environmental impact assessment; but they did not grant locals veto power to stop the siting of a project. Although the conflicts were brought to the courts in Shibuya and Toshima, the judicial system did not favour local opposition either. Generally speaking, it is difficult, if not impossible, to stop governmental projects by the judicial system in Japan (Kajiyama, 2004). As for waste disposal facilities in the 23 wards of Tokyo, there are cases which were brought into the court; the Yaguchi plant in Ota ward in the 1950s, the Kita plant in the 1960s, the Suginami plant in the 1970s, the Toshima plant in the 1990s, and the Shibuya plant in the 2000s. However, none of them were stopped and the incinerators were built eventually. Most of the opposition movements, which initially demanded the cancellation of the projects, soon turned to negotiation on the terms and conditions as the construction was proceeding.

It is true that, in the 1970s, the persistent local opposition, especially the one in Suginami, prevented the overall progress of the incinerator projects and led to the decline of IWWD. The power of the resistance in Suginami during the 1970s was attributable to the ownership over the concerned land. In contrast, the TMG avoided lands which were owned by many individuals in the 1991 siting plan. Learning the lesson from the Suginami project in which the land was owned by many landlords, the TMG selected public lands or private lands owned by corporations (Shinjuku-ku-gikai, 1992). Actually, half of the proposed sites in the 1991 siting plan were owned by the central government[7] or by the TMG.[8] Although the others were private lands, they were owned not by local residents, but by private enterprises.[9] The land purchases were generally agreed upon before the announcement of the plan, although the prices and other conditions were still to be negotiated.[10] Not being in possession of the lands for the proposed sites, the opposition movements were not strong enough to delay the projects.

This does not mean that stopping an incinerator project was totally impossible; it could not be successful if a ward government and/or a ward council disagreed with it. In fact, there were two incinerator projects in the 1991 siting plan which were cancelled because of local opposition: the project in Setagaya and the incinerator vessel harboured at the existing landfill next to Koto ward. In both of them, the ward government and/or council protested against the project. As a result, they were shelved and ultimately cancelled in the 1997 revision.

Nonetheless, the local opposition and the resultant cancellations of these two projects did not negatively affect the prominence of IWWD. As noted above, these two projects were planned as exceptions to the principle of IWWD, in order to achieve AWI immediately. As the two projects had been stuck in a deadlock, the TMG had to move forward the schedules for the other projects to achieve AWI by the time the new landfill started operation. Thus, these local protests accelerated the incinerator siting based on IWWD rather than hindering it.

To sum up, the slowdown of incinerator siting in the 1997 revision was neither because of scarce land space for the projects nor the local opposition against them. The three projects in Shinjuku, Nakano, and Arakawa were postponed even though the sites were secured for them. Despite the prevalence of local protests, most of them were not able to stop the projects without support from the ward governments and/or councils. Although the two projects were cancelled as the ward governments and/or councils opposed them, these cancellations accelerated the progress of OWOI rather than delaying it.

Prolonged recession and decreasing waste

To explain the slowdown of IWWD in the 1997 revision, it is necessary to take into account the financial crisis of the TMG and the decreasing amount of waste. The bubble economy, which boomed in the late 1980s, burst in 1989, and Japan fell into a long recession. This economic downturn caused the deterioration of the finance of the TMG and the reduction of garbage. As explained in the previous chapter, the effect of this economic change on the siting policies did not surface due to the expectation that the economy would recover soon and waste would keep increasing in the future. However, the impact of the bubble burst gradually undermined the foundation of IWWD and finally led to the siting revision in 1997.

The prolonged economic recession led to the rapid deterioration of the TMG's financial condition. Contrary to the optimistic belief in 1991 that the economy of Tokyo would grow by more than 4% annually until 2000 (Tōkyō-to Kikaku Shingi Shitsu, 1990), the economy suffered a long recession instead. The annual economic growth rate in Tokyo recorded −0.1% between 1990 and 1995.[11] This economic downturn dealt a blow to the finance of the TMG. As noted in the fourth chapter, the budget of the TMG was susceptible to economic fluctuations. The tax revenue of Tokyo was reduced by nearly one trillion yen in the three years from 1992 to 1994 (Nakamura, 1998). It was estimated that there would be a shortage of around 500 billion yen in the budget in 1997, while the reserve funds, which had been accumulated in the 1980s, had dried up (Tōkyō-to Gyōzaisei Kaikaku Suishin Hombu, 1996).

The TMG had to cut down its investments in infrastructure to replenish its coffers. Its investment in 1998 was reduced to 70% of that in 1996, which was around half of what it was in 1992 (Nakamura, 1998). The incinerator projects were no longer an exception. The TMG could no longer afford as many incinerators as IWWD required under this financial crisis. The revision of the 1991 siting plan was written in the Finance Reconstruction Implementation Plan in 1997 (Zaisei Kenzenka Jisshi Iinkai, 1997). The TMG started withdrawing from IWWD in part because the financial turmoil undercut the TMG's capability of putting the OWOI policy into practice.

Furthermore, the actual amount of waste kept decreasing beyond the estimation made in 1991 and thereby diminished the necessity for new incinerators. The amount of waste generated in Tokyo was more sensitive to economic fluctuation than other municipalities because waste from business activities accounted for a larger portion of the total waste generation in Tokyo than it did in other municipalities. The 1991 siting plan was based on the assumption that the amount of waste would gradually increase in the foreseeable future (Tōkyō-to Seisō Kyoku, 1991). However, after it reached a peak of 4.90 million tonnes in 1989, the quantity kept decreasing to 4.13 million tonnes in 1996, an amount much less than the anticipated 4.99 million tonnes in the 1991 siting plan (Tōkyō Nijūsan-ku Seisō Ichibu Jimu Kumiai, 2006).

Moreover, the future amount of waste was anticipated to decline further, taking into account the downward revision of Tokyo's prospected economic growth. Although the TMG still expected that the economy would start to recover from 1997, the growth was estimated to be only 2.25% from 1996 to 2000, and 2.75% from 2001 to 2005 (Tōkyō-to Seisaku Hōdō Shitsu Keikaku Bu, 1997). Thus, future waste production growth was expected to be similarly constrained.

Furthermore, the advancement of waste reduction policies was expected. The siting policy was revised based on the report by the Tokyo Waste Management Advisory Committee in 1997 (Tōkyō-to Seisō Shingikai, 1997). The report strongly recommended a shift to a more sustainable socio-economic system from the one based on mass-production, mass-consumption and mass-disposal. While the report estimated that the amount of potential waste would increase with the growth of real domestic production in Tokyo, this potential increase was expected to be set off by the further advancement in waste reduction policies.

As a result, the report predicted that the amount of waste to be disposed of would keep declining from 4.13 million tonnes in 1996 to 3.80 million tonnes in 2000, and down to 3.65 million tonnes in 2006 (Tōkyō-to Seisō Kyoku, 1997c). Even though the amount was expected to increase from 2006, the growth rate was estimated to be not so high in the long term. The 1991 siting plan was based on the assumption that the amount of waste to be disposed of would be more than 5.00 million tonnes after 2000. The gap between the two estimations was obvious. As the more pessimistic view on the economy and the advancement of waste reduction policies were taken into consideration, the amount of waste in the future was expected to decrease in the medium term and to not increase so much in the long run.

The decline in the actual and the prospected amount of waste made IWWD less convincing to the TMG. As noted, the priority for the TMG was to achieve AWI, in order to secure landfilling space as the final destination of the garbage. However, as the actual amount of waste decreased beyond the expectation, AWI was achieved for the first time when the Edogawa plant, which had been under renovation, resumed operation in February 1997 (Tōkyō-to Seisō Kyoku, 2000). As the amount of waste fell below that in the early 1980s, the goal was attained before any of the new incinerators planned in 1991 were brought into operation. There were three urgent incinerator projects in the 1991 siting plan in order to achieve AWI as soon as possible: the third plant in Setagaya, the incinerator vessel at the existing landfill, and the Sumida plant. However, the first two projects were cancelled and the Sumida plant was still under construction in 1997. Nonetheless, AWI was expected to be accomplished with a 30% margin without building incinerators in Chiyoa, Bunkyo, Taito, Shinjuku, Nakano, and Arakawa wards.

The incineration capacity was recognised as sufficient, even when taking into account the planned reduction in the capacity of some of the existing incinerators. After the Guideline for Dioxin Prevention in 1990 (the old guideline), the Ministry of Welfare issued the Guideline for Dioxin Prevention on Waste Disposal in 1997 (the new guideline). Responding to the tightened regulation on dioxin emission, the TMG planned to renovate six old incinerators built before 1975.[12] Their capacity was to be reduced because larger space was needed to install the latest pollution prevention equipments. The expected capacity reduction amounted to 3,100 tonnes/day[13] in total, which was equivalent to five incinerators as large as the Suginami plant.[14] In spite of this capacity reduction, on top of the cancellation of the two projects, AWI was expected to be achieved without constructing an incinerator in every ward.

Thus, the decreasing amount of waste, resulting from the prolonged economic recession and the rising influence of SWM as the new policy paradigm, made new incinerators increasingly unnecessary and led to the postponement of the projects in the six wards. The idea of siting incinerators in every ward became less cognitively legitimate as the need to construct more incinerators was declining.

Weakened interest of Koto

In addition, Koto ward was losing its motivation to adhere to IWWD. As explained in the previous chapter, the 1991 incinerators siting plan was formulated in response to pressure from Koto ward. The ward requested siting incinerators in every ward in its campaign against the new landfill siting. However, the pressure from the ward was alleviated in the latter half of the 1990s.

Actually, Koto ward did not make a strong protest against the revision of the incinerator siting plan in 1997. When the TMG explained to the ward about the revision, the ward council criticised the revision for interrupting the steady progress of IWWD, which was promised when the ward agreed to the new landfill siting. However, the council accepted the revision with requests for the establishment of AWI, the cancellation of the incinerator vessel project, and the imposition of a certain responsibility for waste management on the wards where the incinerator projects were virtually shelved (Kōwan Rinkai Taisaku Tokubetsu Iinkai, 1997).

Similarly, Koto ward did not persist in its protest when the ward was asked for an agreement over the start of the B block of the landfill in 1998 (Kōwan Rinkai Taisaku Tokubetsu Iinkai, 1998). The new landfill siting was settled in 1995 when Governor Aoshima visited the ward and apologised for the disproportionate burden which the ward had carried. To persuade Koto ward, the TMG accepted the block-by-block negotiation in which consultation and agreement with the ward would be required every time a new block would start operation. However, when asked for its agreement on starting B block, the ward accepted it even though the ward could have withheld its consent. Koto ward no longer adhered to the goal of constructing an incinerator in every ward, as compared to the previous period in which the ward persistently pressured the TMG and the other wards to complete the incinerator projects.

This was because the ward became less interested in IWWD as AWI was achieved in 1997 and all waste for incineration was expected to be incinerated with certainty in the foreseeable future without constructing incinerators in every ward. This does not mean that Koto ward was satisfied with the distribution of the burden of waste disposal; a large number of garbage trucks driving to the new landfill still passed through the ward. However, the burden from the landfill would no longer be reduced by constructing more incinerators once AWI was achieved. Although the incineration capacity in the ward was still the largest among the 23 wards, its interest in siting incinerators in every ward to redress the distributive injustice became less strong. Consequently, the ward started finding another way to rectify the injustice, as explained later in this chapter.

Decline of the self-responsibility of each ward

Weakening influence of IWWD on the devolution

The influence of IWWD on the devolution also started diminishing along with its influence on the incinerator siting. The shift to the self-sufficient incineration performed by each ward was further delayed. While the legal responsibility for waste disposal was to be devolved to each ward in 2000 by the amendment of the Local Autonomy Act in 1998, an issue arose as to what was an appropriate institutional system to perform the incineration. As noted in the previous chapter, the 1994 agreement between the TMG and the wards involved the block incineration system, in which the 23 wards would be divided into several blocks and incineration would be performed block by block. This was supposed to be the transitional measure until enough incinerators for IWWD were built.

However, this policy was retrogressively modified in the Proposal to the Devolution of Waste Management[15] in 1998. The TMG, the union and the 23 wards agreed to adopt the regional incineration system until 2005, in which incineration would be administrated jointly by the 23 wards by forming a local government association. Although the shift to self-sufficient incineration by each ward was still scheduled, with the block incineration system as a transitional measure, the realisation of it was postponed further.

Autonomy system reform in 1998

The decline of IWWD in incinerator siting impacted its influence on the devolution. Responding to the revision of the siting plan in 1997, the union claimed that the devolution must be shelved as the revised siting plan breached the condition for the devolution agreed between the TMG and the union. Although the plan still supported IWWD in principle, argued the union, the achievement of the idea was abandoned as no clear roadmap for IWWD was shown in the policy, leaving OWOI unaccomplished for at least 20 years (Tōkyō Seisō Rōdōkumiai, 1997b). The union concluded that the TMG was not willing to go beyond the block incineration system and insisted that the devolution must be cancelled if IWWD were abandoned.

From the beginning, the union had preferred the regional disposal system, i.e. the status quo, to the local disposal which IWWD required. The union nonetheless demanded IWWD be accomplished, as a condition for the devolution, to prevent this institutional reform from being achieved. Furthermore, the union thought that IWWD became even more impracticable as siting incinerators in every ward would lead to excessive incineration capacity and argued that a new way of waste management should be considered, based on a regional perspective rather than IWWD (Tōkyō Seisō Rōdōkumiai, 1997a). Thus, the strategy of the union, which had once engendered the political driving force for IWWD, now worked against the idea as the siting plan was revised.

The union tried to stop the devolution and maintain a regional waste management system. To amend the Local Autonomy Act and attain autonomy reform, the TMG had to persuade the union, as the Ministry of Home Affairs (MOH) had insisted that the concerned parties must come to an agreement (Miyake, 2006). The TMG proposed to the union the postponement of the date of the devolution, with which waste management would be performed regionally by the TMG until a day provided by a law in the future. Later, the TMG suggested devolving the legal responsibility of waste management to the 23 wards but entrusting the actual services to the TMG to be performed regionally. Although the union agreed to these proposals, which were intended to virtually cancel the devolution, both of them were rejected by the MOH, who maintained that the devolution should be achieved both in name and in reality, if the 23 wards were to become basic local municipalities (Tōkyō Seisō Rōdōkumiai, 1999).

On the other hand, the 23 wards were seeking a way around the agreement of the union to pull off the autonomy system reform. Previously, the 23 wards' strategy relied on the political influence of Governor Suzuki on the MOH as a former administrative vice minister of the ministry. However, this strategy failed when Suzuki stepped down in 1995, and his successor, Nobuo Ishihara, who was also a former administrative vice minister of the MOH, lost the election to independent Yukio Aoshima. As the negotiation with the union and the ministry ran into a deadlock, the 23 wards sought an alternative route to pull off the reform (Miyake, 2006); they tried to persuade the ministry by making use of their influence on and connections with representatives in the Tokyo Metropolitan Assembly and the National Diet[16] through the Liberal Democratic Party of Tokyo. This strategy succeeded and the ministry agreed to amend the act without the consent of the union. The request was officially made by the TMG and the 23 wards to the ministry and the act was amended in May 1998. The legal responsibility of waste management was to be devolved to each ward in 2000.

The further postponement of self-sufficient incineration

The fight over IWWD continued even after the amendment of the Local Autonomy Act. It was obvious that self-sufficient incineration was impossible for the time being, because it would take another 20 years to achieve OWOI, even if the incinerator siting proceeded on schedule as planned in the Tokyo Slim Plan 21. Consequently, an argument arose as to what would be the appropriate administrative form to perform incineration, while the legal liability of waste management from collection to final disposal was to be devolved to each ward in 2000. The battle further delayed the realisation of the self-responsibility of each ward in waste disposal.

The union argued that all waste management services, ranging from collection and transportation to final disposal, should be performed by one regional governmental organisation, such as a local government association or a local government regional coalition.[17] Now that the once-promised OWOI was withdrawn, argued the union, waste management should be performed regionally (Tōkyō Seisō Rōdōkumiai, 1997a). When the TMG and the 23 wards came to an agreement and made the official request to the central government at the end of 1997 without the consent of the union, the TMG promised to take a regional approach after the devolution to appease its anger; the union demanded the TMG keep this promise.

To keep this promise to the union, the TMG asked the wards to adopt some regional systems for all the waste services including incineration. The 23 wards, however, at first insisted on taking the block incineration system as a transitional measure for realising self-sufficient disposal by each ward, as scheduled in the 1994 agreement. In particular, the wards which persuaded locals to accept incinerators to further the case of autonomy reform and IWWD during the 1990s were reluctant to adopt a regional incineration system. The council of Koto ward was also concerned that the burden on the ward would continue if incineration were to be administrated regionally (Kusei Taisaku Tokubetsu Iinkai & Seisō Kōwan Rinkai Taisaku Tokubetsu Iinkai, 1998).

Nonetheless, they agreed on adopting the regional disposal system for the time being instead of shifting to the block incineration system immediately. One reason was the temporary reduction of the incineration capacity due to dioxin prevention. As already noted earlier in this chapter, some of the plants would have to be renovated to put in place dioxin prevention mechanisms, according to the new regulation which was imposed in 1997. Until 2005, some of the incinerators would stop operation in turn to undergo this renovation, thereby decreasing the total incineration capacity to just below 110% of the quantity of waste for incineration (Seisō Ikan Jisshi Iinkai, 1998). Given this temporary reduction of incineration capacity, AWI was considered difficult to keep achieving even under the block disposal system, as AWI under the block disposal system required larger incineration capacity than under the regional disposal system. After examining the impact of the dioxin prevention measures on incineration capacity, the 23 wards decided to adopt the regional disposal system by forming a local government association until the renovation was completed in 2005, although the 23 wards would adhere to performing collection and transportation individually (Tokubetsu-ku-chō Kai, 1998). Although the shift to IWWD, with block incineration as a transitional measure, was still scheduled to follow, the realisation of self-sufficient incineration by each ward was further postponed.

Furthermore, the 23 wards did not seriously advocate the self-responsibility of each ward for incineration in the first place. In the history of the autonomy expansion movement, they had demanded the devolution of garbage collection and transportation, but not its disposal. In the previous period, nonetheless, they had to accept IWWD in order to persuade the union for the sake of the autonomy expansion reform. Even the mayor of Koto ward did not welcome IWWD being made a part of the devolution, suspecting that the idea was exploited by the union to frustrate the reform (Murohashi, 1998). When the autonomy expansion reform was attained, there was no imminent political reason for them to maintain the self-responsibility of each ward for its incineration.

In the end, none of the major actors supported IWWD seriously. What made the self-responsibility of each ward for incineration an influential idea was the union's strategic intention to prevent the devolution; the union's interest had been in keeping the regional disposal system rather than IWWD from the beginning. The TMG and the 23 wards accepted IWWD for the devolution because they needed to persuade the union in order to attain their political goal. While waste management was finally devolved to the wards, the idea that each ward should perform incineration self-sufficiently was losing its prominence in the relevant policies.

Abandonment of IWWD and a move towards a new way for distributive justice

Abandonment in 2003

The influence of IWWD continued to decline after 1997 and this idea of distributive justice was finally abandoned in 2003. The abandonment of IWWD started with the cancellation of the incinerator projects (Figure 6.2).

The 1991 plan intended to site incinerators in every ward, although this was not explicitly set out for three wards, Bunkyo, Taito and Shinjuku. When the siting plan was revised in 1997, five projects were under progress: Chuo, Minato, Sumida, Shibuya and Toshima. As a result, there were 21 incinerators located in 17 out of the 23 wards[18] as of 2003, while the projects in the other six wards were postponed in the 1997 revision.

The Tokyo Slim Plan 21, which was a revision of the 1991 siting plan, was taken over by the Clean Association of Tokyo 23 Wards,[19] the local association formed by the 23 wards to perform incineration regionally after the devolution in 2000. Although the Municipal Waste Management Plan in 2000[20] still scheduled the construction of incinerators in Shinjuku, Nakano, and Arakawa, the association started reconsidering these projects in 2001. In 2002, the association reported that no more incinerators were necessary (Tōkyō Nijūsan-ku Seisō Ichibu Jimu Kumiai, 2002). Receiving this report, the 23 wards began re-examining the appropriateness of OWOI and the shift to self-sufficient incineration by each ward. In 2003, the mayors of the 23 wards came to the conclusion that new incinerators were no longer necessary (Tokubetsu-ku-chō Kai, 2003b) and cancelled the projects in the six wards as shown in Table 6.2.

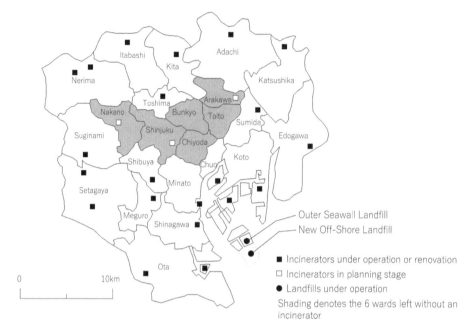

Figure 6.2   Location of incinerators and landfills in 2003.

Adapted from a map of Tokyo by CraftMap (URL: http://www.craftmap.box-i.net/) based on Tōkyō Nijūsan-ku Seisō Ichibu Jimu Kumiai (2002, 2006).

Table 6.2 Incinerator projects in the 1997 and the 2003 revision

Projects	1997 revision		2003 revision
	Construction	Operation	
Sumida	started in 1994	1997	under operation
Setagaya	cancelled		
Incinerator vessel	cancelled		
Minato	started in 1995	1998	under operation
Toshima	started in 1995	1999	under operation
Shibuya	1998	2001	under operation
Chuo	1998	2001	under operation
Chiyoda	2015 (site not specified)	2018	cancelled
Nakano	2010	2013	cancelled
Arakawa	2010	2013	cancelled
Shinjuku	2010	2013	cancelled
Taito	2012 (site not specified)	2015	cancelled
Bunkyo	2012 (site not specified)	2015	cancelled

Source: Tōkyō-to Seisō Kyoku (1997c) and Tokubetsu-ku-chō Kai (2003b)

This abandonment of the OWOI policy immediately led to the abandonment of the self-responsibility of each ward for incineration. After waste management was devolved to the wards, they started examining the shift from regional incineration to block incineration and to self-sufficient incineration as planned in 1998. As a result, they concluded that the 23 wards would jointly be responsible for the incineration and cooperatively perform the disposal, renouncing the agreements made in 1994 which set out the roadmap to achieve the self-sufficient incineration system (Tokubetsu-ku-chō Kai, 2003b). The 23 wards decided to continue the regional incineration system by the cleaning association (Tokubetsu-ku-chō Kai, 2003a; Tokubetsu-ku-chō Kai Jichi Kenkyū Dai Ni Bunka Kai, 2003); the shift to self-sufficient incineration was abandoned. IWWD lost its influence in terms of both the siting of incinerators and the responsibility of each ward to incinerate its own waste.

Decreasing waste and financial difficulty

One of the causes behind this further decline of IWWD was the decreasing amount of waste, beyond even the estimation made in 1997. Figure 6.3 indicates the actual amount of waste and the estimations made in 1991, 1997 and 2002. The Tokyo Slim Plan 21 in 1997 was based on the estimation that the amount of waste would decline from 4.13 million tonnes in 1996 to 3.80 tonnes in 2000, taking into account the economic recession and the advancement of waste reduction policies.

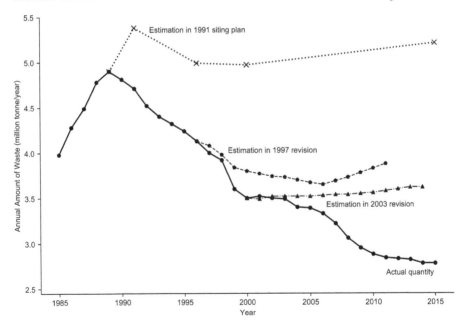

Figure 6.3 Actual quantity of waste and waste growth estimations in 1991, 1997 and 2002.

Source: Tōkyō Nijūsan-ku Seisō Ichibu Jimu Kumiai (2002, 2006, 2013) and Tōkyō-to Seisō Kyoku (1991, 1997c, 2000)

However, the actual amount of waste was already reduced to 3.50 million tonnes by 2000.

The economic recovery was slower than expected. While the 1997 plan was based on the assumption that the economy would grow annually by 2.25% from 1996 to 2000, and by 2.75% from 2001 to 2005, the average annual economic growth rate was 1.7% from 1996 to 2000 and 0.6% from 2001 to 2003.[21] Given this slow economic recovery, the revision estimated around 2% annual economic growth until 2015.[22]

The further advancement of waste reduction policies was also anticipated. SWM was becoming increasingly influential in Japan and being institutionalised into legislations from the latter half of the 1990s to the early 2000s. The Fundamental Law for Establishing a Sound Material-Cycle Society,[23] which institutionalised the 3Rs and expanded producer responsibility, was enacted in 2000. New legislations on recycling of containers and packages, home appliances, food, and so forth, were put into effect one after another during this period. While the 2% annual economic growth until 2015 might potentially increase the amount of waste, this potential increase was expected to be offset by further advancement in waste reduction policies. As a result, the association made the estimation in 2002 that the quantity of waste would remain almost constant for the foreseeable future (Tōkyō Nijūsan-ku Seisō Ichibu Jimu Kumiai, 2002), in contrast to the 1997 siting plan, which had anticipated that waste would start growing from 2006.

This downward trend of waste growth widened the gap between IWWD and AWI in the 23 wards as a whole. AWI had been achieved since 1997; there was already sufficient incineration capacity to continually incinerate all waste to be incinerated. Rather, given the decreasing amount of waste, the existing incineration capacity was already excessive. Even though the capacity of the six incinerators was reduced due to dioxin prevention measures, the total capacity of the 21 incinerators in the 17 wards amounted to 3.67 million tonnes/year (Tōkyō-to Seisō Kyoku, 1997a), while the amount of waste for incineration was only 2.86 million tonnes in 2000 (Tōkyō Nijūsan-ku Seisō Ichibu Jimu Kumiai, 2013) and was not expected to increase in the future; new incinerators were not necessary in regard to AWI in the 23 wards as a whole.

This then raised a dilemma between IWWD and AWI. When waste management was devolved to the 23 wards, they still assumed that each ward, as a basic local government, should perform incineration self-sufficiently in the future. Although the 23 wards as a whole had sufficient incineration capacity necessary for AWI, they needed to construct new incinerators in the wards without one if they were to aim for self-sufficient incineration by each ward. They had to decide whether to pursue IWWD even when the incineration capacity was more than enough for AWI in the 23 wards as a whole.

Moreover, there were also financial concerns. The finances of the 23 wards had been deteriorating since 1992 (Inoue, 2009), as the economy did not recover even after the turn of the century. This poor financial condition made the wards negative to the three projects in Shinjuku, Nakano and Arakawa. As their total cost was estimated to be around 110 billion yen including land purchase (Tōkyō

Nijūsan-ku Seisō Ichibu Jimu Kumiai, 2002), the 23 wards had second thoughts on whether to proceed with the projects as planned.

To make matters worse, there was little chance to gain the subsidy for the projects from the central government. The central government was also experiencing a financial crisis and consequently cutting subsidies for local governments. In addition, the Ministry of Health and Welfare, which had the jurisdiction over waste management back then,[24] encouraged the idea of several local municipalities jointly utilising a small number of large incinerators, to prevent dioxin pollution from incinerators. To restrain dioxin production, waste needed to be incinerated steadily and continually at a high temperature, which was difficult with small incinerators. As such, the ministry encouraged the decommissioning of small incinerators and supported regional disposal with large incinerators shared by several local municipalities forming local government associations or regional coalitions. Although all of the incinerators in the 23 wards were large enough and met the criteria issued by the ministry, the subsidy for the new projects was not likely to be approved when the ministry was driving for a regional, concentrated incineration system and there was already sufficient incineration capacity in the 23 wards as a whole (Tokubetsu-ku-chō Kai Jimu Kyoku, 2006; Tōkyō Nijūsan-ku Seisō Ichibu Jimu Kumiai, 2002). Without the subsidy from the national government, the 23 wards had to pay for the cost of the incinerator projects by themselves.

Disagreement among the wards

Given the decreasing amount of waste and the ongoing financial difficulty, the 23 wards had to decide whether or not to cancel the three projects.[25] Although the 23 wards had been united in advocating IWWD through the political battle for autonomy expansion in the 1990s, the resultant devolution ironically exposed disharmony among them. In this argument on whether to proceed with OWOI, eight wards argued for the projects and ten against, while the rest were undecided (Shinjuku-ku-gikai, 2002a).

Those who insisted on new incinerators were the very three wards with projects under way, Shinjuku, Nakano, and Arakawa, as well as those who made efforts to site incinerators in the 1990s.[26] The three wards requested the association and the other wards to purchase the lands as planned (Nakano-ku, 2001; Shinjuku-ku, 2001; Shinjuku-ku-gikai, 2002b). They thought that the sites were the only lands available for incinerators in the three wards and it would be impossible for them to have incinerators if this opportunity was missed.

They feared that not having an incinerator could make waste disposal in the wards insecure. Before the devolution, the TMG had been the one responsible for incineration in the 23 wards of Tokyo; disposal of waste in wards without incinerators was ensured in this regional incineration system. However, the legal responsibility of waste management was devolved to the wards in 2000. Moreover, the shift to an independent, self-sufficient incineration system in the future was still scheduled. It was also anticipated that the ownership of the incinerators would be transferred from the cleaning association to each ward where they were located. This situation made the wards without an incinerator afraid that their waste might be refused by other wards and be stranded if they were left without an incinerator in the self-sufficient waste disposal system (Tokubetsu-ku-chō Kai Jimu Kyoku, 2006). This anxiety was strong especially for Shinjuku, which was the target of the garbage blockade by Koto ward at the end of 1991 (Shinjuku-ku-gikai, 2002b). As IWWD was influential among the 23 wards and the neighbours around the incinerators, argued Shinjuku ward, some wards and neighbours seemed reluctant to accept waste outside of the ward.

In fact, several incinerators, such as the Suginami plant, the Hikarigaoka plant in Nerima ward, the Meguro plant, and the Minato plant, had imposed certain conditions to control waste outside of the ward coming into the plants, such as accepting only the same amount of waste as the waste generated inside the ward or simply restricting out-of-the-ward garbage (Kōtō-ku Rinkai Taisaku Shitsu, 1999; Tokubetsu-ku-chō Kai Jimu Kyoku, 2006). In addition, some of the large plants, which had accepted waste outside of their own ward, were downsized due to dioxin prevention measures. Given this situation, the three wards wanted a guarantee that their waste would not be refused: they called for the shift to the self-sufficient incineration to be renounced if the land purchases were to be cancelled.

On the other hand, other wards were reluctant to pay for the incinerators given their financial situation. Before the devolution, the 23 wards did not have to pay for the incinerators, at least directly, because the TMG had been the one who paid for the incinerators. However, after the administrative responsibility was devolved to the 23 wards in 2000, they had to pay for the new projects by themselves. As they chose the joint regional incineration system until 2005 by forming a local government association, the cost of the incinerators was to be shared by the 23 wards. Without the subsidy from the central government, each ward had to bear around 5 billion yen on average for the projects. Some of the wards, especially the ones which had enough incinerators to deal with their own waste, were unwilling to take on the cost of the incinerators for the other three wards.

Thus, the diverging interests among the 23 wards surfaced, given the declining necessity for the new incinerators, the severe financial condition, and the institutional change as a result of the autonomy reform. Although the disagreement between the haves and the have-nots had existed below the surface even in the 1990s, they were united for the sake of the autonomy reform. Once this common goal was attained, however, the disharmony in relation to IWWD surfaced.

Even in Koto ward, there were doubts about the idea of siting incinerators in every ward. In the ward council, which had been the spearhead of the campaign for this idea of distributive justice, some started arguing against the new incinerators siting.[27] It is true that there was concern that the cancellation of the projects would lead to the failure of IWWD, which had been the ideational foundation of its claims for distributive justice; the ward could not easily backtrack from the idea which it had advocated since the early 1970s. However, the ward was reluctant to bear the cost of the incinerator projects under its severe financial condition.

Furthermore, even in Koto ward, it was argued that OWOI was an out-of-date idea and more effort should be placed on the reduction and recycling of waste to decrease the number of incinerators rather than increase it. The recognition of the idea as a solution to the garbage problem was being undermined even in its birth-place. As a result, the mayor of the ward rather took the initiative of cancelling the three projects.[28]

This argument over IWWD was settled by renouncing both OWOI and the shift to self-sufficient incineration in each ward. In July 2003, the Association of Ward Mayors confirmed that the new incinerators were no longer necessary, given the declining amount of waste production and the serious financial condition, and decided that disposal would be carried out regionally by the joint responsibility of the 23 wards as a whole, regardless of whether each ward had an incinerator or not (Tokubetsu-ku-chō Kai, 2003b). The three incinerator projects were then cancelled. The 23 wards also abandoned the 1994 agreement which showed the roadmap from the regional disposal to self-sufficient disposal via block incin-eration as the transitional system. The self-sufficient, independent incineration, which was once promised through negotiation over the devolution, was given up together with the realisation of OWOI.

Towards a new scheme of distributive justice

Although it was confirmed that the incineration system would be operated in a cooperative and regional manner, distributive injustice still existed. As IWWD was abandoned, a new scheme for distributive justice was required. It was pointed out that an imbalance existed in the distribution of incinerators and their capac-ity (Tokubetsu-ku-chō Kai Jichi Kenkyū Dai Ni Bunka Kai, 2003). While the six wards, i.e. Chiyoda, Bunkyo, Taito, Shinjuku, Nakano, and Arakawa, were left without an incinerator, there were 21 incinerators in 17 out of the 23 wards. Nevertheless, the incineration capacity among these 17 wards ranged from 200 tonnes/day in Shibuya ward to 2,200 tonnes/day in Koto ward. Since the first garbage war in the 1970s, IWWD had been regarded as the fundamental way to realise distributive justice among the 23 wards by siting incinerators in every ward. As the necessity to construct new incinerators was rejected, they needed a different scheme to redress the remaining inequality.

Koto ward took the initiative to redress the injustice. Koto ward claimed that a disproportionate burden was still being carried by the ward. There were two incinerators in Koto with 2,200 tonnes/day capacity, which incinerated 21% of the waste for incineration being generated in the 23 wards (Kōtō-ku-gikai, 2005). Of the waste incinerated in Koto, 74% came from outside of the ward. As the other big incinerators such as the Adachi plant, the Katsushika plant, and the Oi plant in Shinagawa ward were scaled down for dioxin prevention measures, the incineration capacity in Koto was remarkably large among the 23 wards.[29] Koto also complained that more than one million garbage trucks came through the ward every year because, in the ward and the reclaimed land next to the ward, there were many waste disposal facilities besides the two incinerators, such as intermediate treatment facilities for waste not for incineration, a bulky garbage treatment facility, and the landfill.

Koto ward once again launched a campaign to redress the imbalance in the burden distribution of waste disposal. Since the abandonment of IWWD in 2003, how to balance the burden of waste disposal facilities had been discussed in a meeting of the deputy mayors of the 23 wards. However, even in 2005, the meeting had not shown any practical scheme. Frustrated with the stagnated discussion, the council of Koto began activities to make the other wards understand the burden that Koto suffered. Starting with making a pamphlet to illuminate the disproportionate burdens that the ward had borne,[30] the Koto council launched a special committee in 2006. The committee sent a letter to the Association of Special Ward Mayors, the Association of Special Ward Chairpersons, the meeting of the deputy mayors, and the cleaning association, and asked them to alleviate the disproportionate burdens on Koto. The letter referred to a "grave determination", which implicitly suggested stopping the operation of an incinerator in the ward (Kōtō-ku-gikai, 2006a).

With increasing protest from Koto, the 23 wards resumed the discussion over the fairness issue at the beginning of 2006. Receiving the letter from Koto ward, the head of the Association of Special Ward Mayors and the cleaning association expressed their understanding and stated that the issue would be considered (Kōtō-ku-gikai, 2006b; Tosei Shimpō, 2006). In October 2006, the deputy mayor of Katsushika ward proposed a draft of a scheme to resolve the disparity, including the levelling of the amount of waste incinerated among the plants, a further reduction of waste, the cancellation of the local pacts which limited the operation of the incinerators, and the introduction of monetary compensation.

A new distributive justice policy was decided upon in 2008 (Tokubetsu-ku-chō Kai Jimu Kyoku, 2009). The new policy was a combination of monetary compensation and a source reduction approach, in which distributive justice was expected to be achieved by reducing the amount of waste produced rather than constructing new incinerators, along with a monetary compensation scheme as a secondary measure. The purpose was to balance the amount of waste incinerated among the 16 wards with incineration capacity enough for their own waste, according to a certain criterion which was defined by the amount of each ward's own waste, plus 15% of the average amount of waste in the 16 wards.[31] However, the 16 wards had to accept more than the amount decided by this criterion because the current amount of waste for incineration in the 23 wards exceeded the total amount of waste to be disposed of in the 16 wards according to the criterion. Therefore, waste reduction by 20% in 10 years was set as a goal to achieve this levelling.

At the same time, a monetary compensation scheme was introduced as a temporary measure. Of the 16 wards, ones that incinerate more than the criterion were compensated in proportion to the excessive amount. The cost for the compensation was divided among the six wards without an incinerator, the ward with an incinerator but not enough capacity for its own waste (Shibuya ward), and any ward among the 16 wards whose incineration amount did not reach the level determined by the criterion.[32]

This new policy was different from IWWD in three points. Firstly, this scheme was based on a joint and regional incineration system, rather than the self-sufficient, independent disposal which IWWD required. The 16 wards with incinerators were supposed to accept waste from other wards on top of their own waste. Even wards that could take care of their own waste had to pay if they incinerated less than the amount of waste determined by the criterion, that is, their own waste plus 15% of the average amount of waste in the 16 wards. This scheme required the wards with incinerators to dispose of more than their own garbage, contrary to IWWD.

Secondly, the new scheme was to lift the local pacts with neighbours which limited the amount of waste being incinerated and/or waste moved across borders. Those local agreements were made as a result of negotiation with the local residents over the incinerators siting. As these incinerators were planned in the name of IWWD, the claims for putting the limitation on their operation were ideationally difficult to refute. However, those pacts were regarded as the obstacle to the cooperative regional disposal and the levelling of the incineration capacity among the wards.

Last, but not least, this new approach was based on different ideas of distributive justice: source reduction and compensated equity. As the 23 wards confirmed that new incinerators were no longer necessary, they had to consider a way to redress the distributive injustice without constructing more incinerators. The central idea of this new scheme was the source reduction approach in which distributive justice was to be achieved by reducing the amount of waste and incineration capacity in contrast to IWWD. IWWD had always been connected with incinerationism and the need to construct more incinerators. When new incinerators were no longer necessary, IWWD gave way to the new approach which was centred on source reduction along with compensation as a supplementary measure.

Conclusion

During this period, the changes in the four variables adversely affected the prominence of IWWD. As a result of the interaction between these changes, the enthusiasm for IWWD evaporated quickly and the idea lost its influence on both the siting of incinerators and the institutional reform of waste management.

The cognitive legitimacy of the idea of siting incinerators in every ward became weaker as the declining amount of waste production made OWOI excessive in relation to the capacity needed for AWI. Even in Koto ward, it was recognised that siting incinerators in every ward to rectify the disparity among the wards was no longer reasonable. The idea of distributive justice also became less congruent with the interests of the major stakeholders. Koto ward became less interested in OWOI as AWI had already been achieved in 1997. While the three wards, i.e. Shinjuku, Nakano, and Arakawa, were afraid of being left without an incinerator, others, including Koto ward, were not willing to pay the cost for them. The power of the TMG and the 23 wards to implement OWOI was undermined as their financial positions were deteriorating. There was little chance of acquiring the subsidy for the new incinerators, given that the incineration capacity on the whole was more than enough for AWI.

These variables changed in relation with the changing exogenous environ-ments. The decline in waste generation, which undermined the recognition of IWWD as problem-solving and made Koto ward less interested in the idea, was mainly due to the prolonged economic recession and the rise of SWM as a new policy paradigm. The economic recession brought about the financial crisis and damaged the governmental capacity to realise IWWD, and then brought forth a disagreement in interests among the 23 wards. The dioxin problem led to a national policy which advocated the implementation of a regional disposal sys-tem. Unfortunately, there was only a slight chance of acquiring a national subsidy for the incinerator projects, as there was more than enough incineration capacity for AWI. The idea of siting incinerators in every ward lost its influence on policy as a result of the interaction between these changes in the variables.

On the other hand, the decline of the idea of siting incinerators in every ward triggered the decline of the self-responsibility of each ward for incineration. The self-responsibility of each ward had been recognised as problem-causing rather than problem-solving throughout the four periods. Paradoxically, it was the interest and power of the union, interacting with the other variables, which engendered the political momentum for IWWD in the early half of the 1990s. When the 1991 sit-ing plan was reconsidered in the 1997 revision and the union insisted that all of the waste management services should be performed regionally, no parties seriously advocated the idea that each ward should perform incineration independently.

As the 23 wards managed to overcome the union's opposition to the devolution and achieved autonomy expansion, they no longer had a strong reason to adhere to the self-responsibility of each ward in waste disposal. When the dioxin crisis as another change in exogenous environments resulted in the tightened regulation and led to the temporary reduction of the incineration capacity, even the shift to the block incineration system was recognised as problematic and was postponed. After the devolution reform, the shift to the self-sufficient incineration system was abandoned, together with the idea of siting incinerators in every ward.

Notes

1 Tōkyō Surimu Puran 21 [東京スリムプラン21] (Tōkyō-to Seisō Kyoku, 1997c). This was a general long-term waste management plan, which was required by the Waste Disposal & Public Cleaning Act.
2 The only exception was Sumida. Although the ward and the ward council once inter-vened the siting process due to issues over conditions for the incinerator, no strong opposition occurred from the neighbours.
3 For instance, see Komatsu (1995), Matsunaga (1996), and Saito et al. (1996).
4 For instance, see Sugaya (1994).
5 For instance, see Haikibutsu o Kangaeru Shimin no Kai (1985–2008), Kajiyama et al. (2006), Meguro Seisō Kōjō no Kensetsu ni Hantai Suru Jimoto Yūshi no Kai (1993), and arguments against incinerators in ward councils in the 1990s.
6 Some of anti-incinerationism movements in the 1990s proposed an alternative to conven-tional incinerators, that is, Refuse Derived Fuel (RDF). RDF, which was made by crush-ing, drying, pressing, and solidifying garbage, attracted attention during this period. Anti-incinerationism advocated a facility which contributed to recycling in place of a

conventional incinerator which just burned waste. Among recycling facilities, a RDF plant was often proposed as a desirable alternative for disposing of waste for incineration, because RDF was believed to be "a dream technology" which could turn garbage into fuel without emitting pollutive substances. A RDF plant was proposed in Minato, Nakano and Toshima (Muto, 1995; Sugaya, 1994; Toshima Shimbun, 1995a, 1995b). In Minato, the ward council requested a RDF plant instead of a conventional incinerator responding to the petition from its citizens (Minato-ku-gikai, 1992). However, a RDF plant was rejected by the TMG, although an experimental plant was accepted in Minato, as the council demanded it. The TMG thought that the technology was still at the experimental stage (Tōkyō-to-gikai, 1994, 1995).

7 Shinjuku and Nakano.
8 Chiyoda, Sumida, and Setagaya.
9 Minato, Shibuya, Toshima, and Arakawa.
10 Responding to a question about the practicability of the incinerator projects, the TMG stated that general agreements had already been made with landowners for all of the private lands (Tōkyō-to-gikai, 1991).
11 Gross prefectural expenditure at constant prices based on Prefectural Accounts 1990–2003 (93SNA, benchmark year=1995).
12 The Oi plant, the Tamagawa plant in Ota ward, the Setagaya plant, the Itabashi plant, the Adachi plant and the Katsushika plant.
13 The incineration capacity was reduced as follows: The Oi plant from 1,200 tonnes/day to 600 tonnes/day; the Tamagawa plant from 600 tonnes/day to 300 tonnes/day; the Setagaya plant from 900 tonnes/day to 300 tonnes/day; the Itabashi plant from 1,200 tonnes/day to 600 tonnes/day; the Adachi plant from 1,000 tonnes/day to 700 tonnes/ day; and the Katsushika plant from 1,200 tonnes/day to 500 tonnes/day (Tōkyō-to Seisō Kyoku, 1997a).
14 The Suginami plant had three kilns, each of which could dispose of 300 tonnes/day, but one of them was a backup and therefore not used.
15 Seisō Jigyō no Ikan ni Kansuru Teian [清掃事業の移管に関する提案] (Tōkyō-to, 1998).
16 Centred in this political scene was the former minister of the Ministry of Home Affairs, Takashi Fukaya (Tōkyō Seisō Rōdōkumiai, 1999). With the introduction of the single-seat electoral district system as a result of the political reform in 1994, the representative of the National Diet needed the support of the local members of the party and had to show more interest in local issues.
17 A local governmental regional coalition consists of multiple local autonomies and jointly administrates governmental services. This institution is similar to a local governmental association, but more flexible and able to accept authorities and services from prefectures and the state.
18 There are two incinerators in Koto, Ota, Setagaya and Nerima.
19 The official English name is the "Clean Authority of Tokyo Waste disposal of Tokyo 23 cities".
20 Ippan Haikibutsu Shori Keikaku [一般廃棄物処理計画] (Tōkyō Nijūsan-ku Seisō Ichibu Jimu Kumiai, 2000).
21 Gross prefectural expenditure at constant prices based on Prefectural Accounts 1996–2009 (93SNA, benchmark year=2000) and Prefectural Accounts 2001–2016 (93SNA, benchmark year=2005).
22 The report cited this figure from the Tōkyō Kōsō 2000 (Tōkyō-to Seisaku Hōdō Shitsu Keikaku Bu, 2000).
23 Junkan Gata Shakai Keisei Suishin Kihon Hō [循環型社会形成推進基本法].
24 The Jurisdiction over waste management was transferred to the Ministry of Environment in 2001.
25 The site in Nakano ward was still owned by the national government. The national government asked the cleaning association to purchase the land without delay. The site in

Shinjuku ward was also rented from the national government, and the association was asked to purchase the land as the rent contract expired in 2003.

26 Although the information on whether each ward voted for or against the projects was not publicised due to the secret nature of the ballots, a rough guess can be made from the records by the councils of each ward. The following arguments are mainly based on the records of ward councils.

27 Kōtō-ku-gikai (2001, 2002, 2003) and interview with a member of Koto ward council (21 November 2011).

28 Interview with an official of Koto ward (18 Novermber 2011).

29 When the TMG proposed in the early 1990s the renovation of the Koto plant, which was the biggest municipal incinerator in Japan with 1,800 tonnes/day capacity, the TMG promised in the Koto council that the capacity of the plant would preferentially be reduced if the 23 wards had more than sufficient incineration capacity (Kōtō-ku-gikai, 2005).

30 Gomi Mondai to Kōtō-ku (Kōtō-ku-gikai, 2005).

31 Shibuya ward was not included because the plant there was too small.

32 The six wards paid in proportion to the amount of its own waste; Shibuya ward in proportion to the amount of waste that was not incinerated on its own; the rest of the 16 wards in proportion to the difference between the amount determined by the criterion and the amount actually incinerated there.

References

Fukuwatari, K. (1992). Komazawa Kōen Nai Seisō Kōjō Kensetsu Keikaku ni Kansuru Seigan.

Haikibutsu o Kangaeru Shimin no Kai. (1985–2008). Haikibutsu Rettō No.4-166.

Ichishima, M. (1992). Shin Setagaya Seisō Kōjō Kensetsu Keikaku ni tsuite Saikentō o Motomeru Chinjō.

Inoue, Y. (2009). Ishihara Tosei to Tokubetsu-ku no Zaisei. *Rubyu Saantoru*, 9, 49–62.

Kajiyama, S. (2004). *Haikibutsu Funsō no Jōzu na Taisho Hō*. Tokyo: Minji Hō Kenkyū Kai.

Kajiyama, S., Hiwatashi, S., Iida, Y., Osaki, J., Ito, K., & Iwasaki, M. (2006). Genkoku Saishū Jumbi Shomen (1).

Komatsu, Y. (1995). Shibuya Chiku Gomi Shōkyakujō Kensetsu Keikaku ni Kansuru Chinjō.

Kōtō-ku Rinkai Taisaku Shitsu. (1999). *Seisō Kōwan Mondai to Kōtō-ku 21*.

Kōtō-ku-gikai. (2001). Kōtō-ku-gikai Seisō Kōwan Rinkaibu Taisaku Tokubetsu Iinkai Kaigiroku: Heisei Jūsannen Jūgatsu Nijūkunichi.

Kōtō-ku-gikai. (2002). Kōtō-ku-gikai Seisō Kōwan Rinkaibu Taisaku Tokubetsu Iinkai Kaigiroku: Heisei Jūyonen Sangatsu Jūgonichi.

Kōtō-ku-gikai. (2003). Kōtō-ku-gikai Seisō Kōwan Rinkaibu Taisaku Tokubetsu Iinkai Kaigiroku: Heisei Jūgonen Jūgatsu Sangatsu Tōka.

Kōtō-ku-gikai. (2005). *Gomi Mondai to Kōtō-ku*.

Kōtō-ku-gikai. (2006a). Tōkyō Nijūsan-ku no Gomi Mondai o Kangaeru Kai: Heisei Jūhachinen Jūgatsu Hatsuka.

Kōtō-ku-gikai. (2006b). Tōkyō Nijūsan-ku no Gomi Mondai o Kangaeru Kai: Heisei Jūhachinen Jūichigatsu Nijūyokka.

Kōwan Rinkai Taisaku Tokubetsu Iinkai. (1997). Shisetsu Seibi no Kangaekata (Soan) oyobi Shigen Junkan Gata Seiō Jigyō eno Henkan ni Muketa Shisaku (Soan) ni tsuite no Kōwan Rinkai Taisaku Tokubetsu Iinkai no Matome.

Kōwan Rinkai Taisaku Tokubetsu Iinkai. (1998). Kōwan Rinkai Taisaku Tokubetsu Iinkai no Matome.

Kusei Taisaku Tokubetsu Iinkai, & Seisō Kōwan Rinkai Taisaku Tokubetsu Iinkai. (1998). Kusei Taisaku, Seisō Kōwan Rinkaibu Taisaku Tokubetsu Iinkai no Matome.

Matsunaga, M. (1996). Shibuya Higashi Chiku Seisō Kōjō Kensetsu Keikaku ni Kansuru Chinjōsho.

Meguro Seisō Kōjō no Kensetsu ni Hantai Suru Jimoto Yūshi no Kai. (1993). *Furimukeba Entotsu.*

Meguro-ku-gikai. (1990). Tōkyō Toritsu Daigaku Iten Atochi ni Kakawaru Seisō Kōjō oyobi Kankei Shisetsu Hantai ni Kansuru Ikensho.

Minato-ku-gikai. (1992). Minato-ku Nai ni Seisō Kōjō o Kensetsu Suru Baai niha Shigenjunkangata Kōjō to Surukoto o Motomeru Yōbōsho.

Mita, T. (1992). Komazawa Orimpikku Kinen Kōen eno Seisō Kōjō Kensetsu ni Hantai shi, Dōkōen no Kakujū o Motomeru Seigan.

Miyake, H. (2006). Nisennen Toku Seido Kaikaku Seisō Jigyō Ikan no Seiritsu to sono Kyōgi Katei. *TIMR Research Paper*, 1, 1–36.

Murohashi, A. (1998). *Waga Machi Kōtō ni Ai to Hokori o.* Tokyo: Sampō Sha Insatsu.

Muto, Y. (1995). Shōkyaku Umetate Gata Seisō Gyōsei kara no Tenkan o Motomete. *Risaikuru Bunka*, 49, 22–29.

Nakamura, Y. (1998). Tōkyō-to ni okeru Zaisei Kenzenka no Torikumi. *Chihō Zaisei*, 37(12), 136–144.

Nakano-ku. (2001). Nakano Chiku Seisō Kōjō Yōchi no Sōki Shutoku ni tsuite.

Saito, S., Takemori, K., Suzuki, T., & Morii, K. (1996). Kōgi Bun.

Seisō Ikan Jisshi Iinkai. (1998). *Chūkan Shori Ikō no Unei Keitai ni tsuite.*

Setagaya-ku-gikai. (1991). Setagaya-ku-gikai Kaigiroku Jūichigatsu Teireikai: Heisei Sannen Jūichigatsu Muika.

Setagaya-ku-gikai. (1992). Setagaya-ku-gikai Kaigiroku Jūichigatsu Teireikai: Heisei Yonen Jūichigatsu Itsuka.

Shinjuku-ku. (2001). Shinjuku Chiku Seisō Kōjō Yōchi no Shutoku ni tsuite.

Shinjuku-ku-gikai. (1992). Shinjuku-ku-gikai Jichiken Kakujū Taisaku Tokubetsu Iinkai: Heisei Yonen Nigatsu Jūhachinichi.

Shinjuku-ku-gikai. (2002a). Shinjuku-ku-gikai Kensetsu Kankyō Iinkai: Heisei Jūyonen Rokugatsu Jūichinichi.

Shinjuku-ku-gikai. (2002b). Shinjuku-ku-gikai Kensetsu Kankyō Iinkai: Heisei Jūyonen Shigatsu Tōka.

Sugaya, H. (1994). Minato Chiku Seisō Kōjō Mondai o Megutte. *Minato Kumin Tsūshin* 1994-1.

Toku Kyōgikai, & Toku Seido Kaikaku Suishin Iinkai. (1994). *Toku Seido Kaikaku ni Kansuru Matome.*

Toku Seido Kaikaku Suishin Iinkai. (1994). *Gutaiteki Kōdō Keikaku no Gaiyō.*

Tokubetsu-ku-chō Kai. (1998). Kihon Hōshin.

Tokubetsu-ku-chō Kai. (2003a). Tokubetsu-ku ni okeru Antei Tekina Chūkan Shori no Arikata ni tsuite.

Tokubetsu-ku-chō Kai. (2003b). Tokubetsu-ku ni okeru Ippan Haikibutsu no Chūkan Shori ni tsuite.

Tokubetsu-ku-chō Kai Jichi Kenkyū Dai Ni Bunka Kai. (2003). *Tokubetsu-ku-chō Kai Jichi Kenkyū Dai Ni Bunka Kai Hōkoku.*

Tokubetsu-ku-chō Kai Jimu Kyoku. (2006). *Seisō Jigyō ni Kansuru Tokubetsu-ku-chō Kai Kentō Shiryō: Heisei Jūsan kara Jūnananendo.*

Tokubetsu-ku-chō Kai Jimu Kyoku. (2009). *Seisō Jigyō ni Kansuru Tokubetsu-ku-chō Kai Kentō Shiryō: Heisei Jūhachi kara Nijūnendo.*

Tōkyō Nijūsan-ku Seisō Ichibu Jimu Kumiai. (2000). *Ippan Haikibutsu Shori Kihon Keikaku.*

Tōkyō Nijūsan-ku Seisō Ichibu Jimu Kumiai. (2002). *Ippan Haikibutsu Shori Kihon Keikaku Minaoshi Kentōkai Hōkoku.*

Tōkyō Nijūsan-ku Seisō Ichibu Jimu Kumiai. (2006). *Ippan Haikibutsu Shori Kihon Keikaku.*

Tōkyō Nijūsan-ku Seisō Ichibu Jimu Kumiai. (2013). *Jigyō Gaiyō Heisei Nijūgonendo Ban.*

Tōkyō Seisō Rōdōkumiai. (1997a). Seisō Jigyō no Kuikan ni Sakidatsu Rōshi Kyōgi no Saishū Tōtatsu Ten to Waga Kumiai no Kenkai ni tsuite.

Tōkyō Seisō Rōdōkumiai. (1997b). "Shisetsu Seibi no Kangaekata" oyobi Shisetsu Seibi ni Kakawaru Genzai no Totōkyoku no Shisei ni Taisuru Waga Kumiai no Kenkai ni tsuite.

Tōkyō Seisō Rōdōkumiai. (1999). *Tōkyō Seisō Rōdōkumiai Gojūnen-shi.*

Tōkyō Seisō Rōdōkumiai, & Tōkyō-to. (1994). Kankeisha toshiteno Tōkyō Seisō Rōdō Kumiai to Totōkyoku no Kōshō Gijiroku no Kakunin oyobi Oboegaki.

Tōkyō-to. (1998). Seisō Jigyō no Ikan ni Kansuru Teian.

Tōkyō-to Gyōzaisei Kaikaku Suishin Hombu. (1996). *Tōkyō-to Zaisei Kenzenka Keikaku.*

Tōkyō-to Kikaku Shingi Shitsu. (1990). *Dai Sanji Tōkyō-to Chōki Keikaku.*

Tōkyō-to Seisaku Hōdō Shitsu Keikaku Bu. (1997). *Seikatsu Toshi Tōkyō Kōsō.*

Tōkyō-to Seisaku Hōdō Shitsu Keikaku Bu. (2000). *Tōkyō Kōsō 2000.*

Tōkyō-to Seisō Kyoku. (1991). *Seisō Kōjō Kensetsu Keikaku.*

Tōkyō-to Seisō Kyoku. (1997a). *Seisō Kōjō Seibi Keikaku (Soan2) ni tsuite.*

Tōkyō-to Seisō Kyoku. (1997b). *Shisetsu Seibi no Kangae Kata (Soan).*

Tōkyō-to Seisō Kyoku. (1997c). *Tōkyō Surimu Puran 21.*

Tōkyō-to Seisō Kyoku. (2000). *Tōkyō-to Seisō Jigyō Hyakunen-shi.* Tokyo: Tōkyō-to Kankyō Seibi Kōsha.

Tōkyō-to Seisō Shingikai. (1990). *Seisō Jigyō no Kongo no Arikata ni tsuite: Chūkan Tōshin.*

Tōkyō-to Seisō Shingikai. (1997). *Seisō Jigyō no Kongo no Arikata ni tsuite: Saishū Tōshin.*

Tōkyō-to-gikai. (1991). Tōkyō-to-gikai Kensetsu Seisō Iinkai: Heisei Sannen Jūichigatsu Jūyokka.

Tōkyō-to-gikai. (1994). Tōkyō-to-gikai Kaigiroku Dai Ikkai Teireikai: Heisei Rokunen Sangatsu Futsuka.

Tōkyō-to-gikai. (1995). Tōkyō-to-gikai Kensetsu Seisō Iinkai: Heisei Nananen Nigatsu Nijūhachinichi.

Tōkyō-to-gikai. (1996). Tōkyō-to-gikai Yosan Tokubetsu Iinkai: Heisei Hachinen Sangatsu Jūyokka.

Tosei Shimpō. (1990). Toritsudai Itenchi e Seisō Kōjō Kensetsu ni Hantai. *Tosei Shimpō Sha,* 29.6.1990.

Tosei Shimpō. (1991). Toritsudai Atochi Riyō Jūgatsusue nimo Ketsuron. *Tosei Shimpō Sha,* 23.8.1991.

Tosei Shimpō. (2006). Nijūsan-ku Gomi Mondai Futan no Kōhei ni Gutai An. *Tosei Shimpō Sha,* 7.11.2006.

Toshima Shimbun. (1995a). Shinise Shōkyakujō ni Hantai. *Toshima Shimbun Sha,* 24.1.1995.

Toshima Shimbun. (1995b). Shōkyaku Gata Seisō Kōjō Kensetsu Keikaku Tekkai o. *Toshima Shimbun Sha,* 21.2.1995.

Zaisei Kenzenka Jisshi Iinkai. (1997). *Tōkyō-to Zaisei Kenzenka Keikaku Jisshi An.*

7   Conclusion

To explain the rise and fall of a particular idea of distributive justice in policies of locally unwanted facility siting, this book analysed the changing influence of IWWD in Tokyo by focusing on the four variables and the interaction between them. To conclude, this chapter summarises the empirical results and discusses the implications.

Summary of the empirical results

This section summarises the results of the case study as outlined in Table 7.1. It shows how each of the four explaining variables affected the dominance of IWWD and then examines how they interacted with one another.

Ideational legitimacy

The ideational legitimacy of IWWD played a significant role in the rise and fall of this idea of distributive justice. As the four empirical chapters of this book clearly showed, the dominance of IWWD in the siting policies was significantly correlated to the degree to which IWWD was considered to be cognitively and/or normatively legitimate (see Table 7.2). For the idea of siting incinerators in every ward, the normative legitimacy had been strong through all of the four periods, while the cognitive legitimacy fluctuated over time. For the self-responsibility of each ward, the normative legitimacy stayed strong while the cognitive legitimacy was always weak. Taken together, the degree of the ideational legitimacy of IWWD was relatively robust in the first and third periods, attracting more support and less opposition. In contrast, it was relatively weak in the second and the fourth periods, drawing less advocacy and more resistance.

The normative legitimacy of the idea of siting incinerators in every ward stayed robust over time due to its congruence with the underlying values which the TMG and the 23 wards held: the efficiency of garbage transportation and the autonomy of each ward, both of which were satisfied through the establishment of a large number of small incinerators evenly distributed among the wards, rather than having a small number of large incinerators concentrated in a few places. In particular, IWWD was normatively compelling in the 1990s as the autonomy

Table 7.1 Changing dominance of IWWD and explaining variables

Dominance of IWWD	First period (1971–1973) Strong but limited	Second period (1974–1989) Weak	Third period (1990–1996) Stronger	Fourth period (1997–2003) Weak
• Siting incinerators in every ward	Strong	Weak	Strong	Weak
• Ideational legitimacy (Nor/Cog)	+/+	+/−	++/+	+/−
• Interests	+	−	+	−
• Power of claimants	+	−	+	−
• Exogenous environments	+	−	+	−
• Institutional responsibility of each ward	Weak	Weak	Strong	Weak
• Ideational legitimacy (Nor/Cog)	+/−	+/−	++/−	+/−
• Interests	−	−	+	−
• Power of claimants	−	−	+	−
• Exogenous environments	−	−	+	−

+: Worked positively for the dominance of IWWD; −: Worked negatively for the dominance of IWWD; Nor: Normative; Cog: Cognitive

expansion movement reached its culmination in this period. This heightened normative resonance with the autonomy of the 23 wards made IWWD undeniable for the TMG and the 23 wards.

In contrast, the cognitive legitimacy of One Ward One Incinerator (OWOI) waxed and waned over time: strong in the first and third periods, but weak in the second and fourth periods. In the first and third periods, IWWD was accepted by the TMG partly because it believed that this conceptual framework provided a simple and clear roadmap to achieve All Waste Incineration (AWI) under the rapidly increasing amount of waste. Although the practicability of siting incinerators in every ward was doubted from the beginning due to limited land availability in some wards, the idea was expected to facilitate incinerators' construction. In the second and fourth periods, however, siting incinerators in every ward was regarded as excessive when compared to the actual capacity needed for AWI due to the slowdown of the waste production.

On the other hand, the idea that a ward should be institutionally responsible for disposal of its own waste was not cognitively convincing to all of the parties throughout the four periods. Devolving waste disposal authority to each ward was thought to worsen the garbage crisis rather than solve it. Because it was considered hardly possible to find a site for incinerators in all of the 23 wards, making each ward institutionally responsible for incineration was recognised as unreasonable. All of the parties, including Koto ward, thought that incineration was better

Table 7.2 Ideational legitimacy and dominance of IWWD

	First period (1971–1973)	Second period (1974–1989)	Third period (1990–1996)	Fourth period (1997–2003)
• Siting incinerators in every ward (Nor/Cog)	+/+	+/–	++/+	+/–
Normative	• Resonance with transportation efficiency and ward autonomy	• Resonance with transportation efficiency and ward autonomy	• Resonance with transportation efficiency and ward autonomy	• Resonance with transportation efficiency and ward autonomy
Cognitive	• Urgent necessity for more incinerators	• OWOI regarded as excessive	• Urgent necessity for more incinerators	• OWOI regarded as excessive
• Institutional responsibility of each ward (Nor/Cog)	+/–	+/–	++/–	+/–
Normative	• Resonance with ward autonomy	• Resonance with ward autonomy	• Resonance with ward autonomy	• Resonance with ward autonomy
Cognitive	• Regarded as impracticable	• Regarded as impracticable	• Regarded as impracticable	• Regarded as impracticable

performed regionally rather than locally. Although the self-responsibility of each ward for waste disposal was normatively irrefutable due to its resonance with the idea of the autonomy of each ward, this lack of cognitive legitimacy made this idea less legitimate overall.

Interests

As Table 7.3 outlines, the degree of the congruence with the interests of the actors constituted a significant part of the story of the rise and fall of this idea of distributive justice. In the first period, the interests for the idea of siting incinerators in every ward were strong, while the idea of devolving the responsibility to each ward was against the interests of the Tokyo Cleaning Workers' Union. This is reflected in the strong but limited influence of IWWD during this period. In the second period, even the interests for the idea of OWOI became weaker, which had a negative impact on the dominance of this idea. In the third period, the interests worked positively for both siting incinerators in every ward and devolving the responsibility to each ward. This congruence with the interests supported the culmination of IWWD in this period. In the fourth period, however, the influence of IWWD fell when the interests for this idea became weaker once again.

In the first period, the idea of siting incinerators in every ward was congruent with, or at least not against, the interests of the major stakeholders (i.e. Koto ward, Tokyo Cleaning Workers' Union, and the rest of the 23 wards). Koto ward, which had long suffered the pollution from the garbage dumping into the coastal landfills, advocated this concept of distributive justice in order to reduce the disproportionate burdens imposed on the ward by facilitating incinerators construction. The idea was congruent with the interests of the union as it recognised that investing in waste disposal facilities would help them to resolve the discrimination against the cleaning workers and to improve their working conditions. The other wards also had interests in facilitating the incinerators' construction to achieve their political goal, i.e. expanding their political autonomy. The idea of siting incinerators in every ward attracted much support in part due to the congruence with the interests of these stakeholders.

In the second period, the interests of the residents and businesses around the proposed sites for the incinerator projects negatively impacted the dominance of IWWD. They worried that the incinerators would bring environmental degradation to their local communities and hinder local developments. The local opposition delayed the implementation of the incinerator siting projects. Furthermore, Koto ward became less interested in the idea as the goal that Koto ward pursued by claiming IWWD was half-fulfilled when the waste growth slowed down and some incinerators planned before the first garbage war were brought into operation.

On the other hand, the self-responsibility of each ward was against the interest of the union. Although the union generally agreed with the idea of siting incinerators in every ward, it intensely opposed any idea that a ward should be institutionally responsible for its own waste disposal. The union feared that the devolution of waste management would weaken the power of the organisation as a whole,

Table 7.3 Interests and dominance of IWWD

	First period (1971–1973)	Second period (1974–1989)	Third period (1990–1996)	Fourth period (1997–2003)
• For siting incinerators in every ward	+ • Koto ward's strong interest • Not against the interests of the 23 wards • Congruent with the union's interest	– • Koto ward's interest weakened • Not against the interests of the 23 wards • Against interest of local residents/businesses	+ • Koto ward's strong interest • Congruence with the interests of the 23 wards	– • Koto ward's interest weakened • Disagreement between the 23 wards
• For institutional responsibility of each ward	– • Against the union's interest	– • Against the union's interest	+ • Against the union's interest but worked positively for IWWD	– • Disagreement between the 23 wards • Against the union's interest

thereby increasing the risk of worsening the working conditions for union members. This interest of the union worked adversely for the idea of the institutional responsibility of each ward during the first and the second periods.

In the third period, the congruence with the interests of stakeholders helped the rise of IWWD. Koto called for IWWD once again to facilitate incinerator construction because un-incinerated waste was still brought to the coastal landfill given the sharp increase of waste in the late 1980s. Having incinerators in their turfs agreed with the interests of the rest of the 23 wards: promoting the devolution of waste management and the autonomy reform. Furthermore, the union, which had intensely opposed the concept of autonomy/self-responsibility of each ward in relation to waste disposal, nevertheless persistently called for the institutional responsibility of each ward in this period. The union's interest was in stopping any devolution of waste management authority to the wards; it tried to prevent the devolution by making IWWD a tough precondition on this institutional reform, although the realisation of IWWD was against its interest. This strategy to prevent the devolution engendered the political driving force for the rise of the idea of the self-responsibility of each ward and thus pushed forward the OWOI policy.

In spite of the enthusiasm in the 1990s, IWWD lost its congruence with the interests of the major actors in the fourth period. Koto ward weakened its claim for IWWD as AWI had been achieved in 1997. The union also stopped calling for IWWD when the strategy to stop the devolution by imposing IWWD failed; it started explicitly arguing against any devolution. The 23 wards were no longer united in advocating IWWD. The conflicting interests among the 23 wards surfaced once the autonomy reform was attained. Shinjuku, Nakano, and Arakawa wards insisted on proceeding with the incinerator siting projects as planned, for they were afraid of being left without an incinerator, since the shift to the self-sufficient incineration system was still scheduled. On the other hand, as the cost of incinerators was to be shared by the 23 wards after the devolution in 2000, some of them, especially those who had incinerators in their wards, were unwilling to pay for the cost of the incinerators in the three wards. As the congruence with the self-interested goals of the stakeholders waned, the prominence of this idea of distributive justice declined in the fourth period.

Power of claimants

The degree of the power of claimants significantly affected the dominance of IWWD as summarised in Table 7.4. In the first period, the rise of the idea of siting incinerators in every ward was backed by the power of Koto ward, while the idea of the self-responsibility of each ward was obstructed by the political influence of the union. In the second period, the inability of the TMG to put the idea into practice brought negative feedback to the dominance of IWWD in the siting policies. Furthermore, the power of Koto ward declined as the landfill crisis subsided. In the third period, the influence of IWWD rose to its peak, with Koto ward and the union as the powerful claimants. The TMG was capable of implementing the OWOI policy, given the relatively weak local opposition and the robust financial

Table 7.4 Power of claimants and dominance of IWWD

	First period (1971–1973)	Second period (1974–1989)	Third period (1990–1996)	Fourth period (1997–2003)
• For siting incinerators in every ward	+ • Koto ward's strong influence	− • Koto ward's influence weakened • TMG not able to implement as planned due to strong local opposition, limited land availability, and financial crisis	+ • Koto ward's strong influence • TMG able to implement as planned due to weak local opposition and robust financial capability	− • Koto ward's influence weakened • TMG and the 23 wards not able to implement as planned due to financial crisis
• For institutional responsibility of each ward	− • Not able to overcome the union's opposition	− • Not able to overcome the union's opposition	+ • The union able to impose IWWD on the devolution	− • The 23 wards able to overcome the union's opposition • The 23 wards' inability to implement the devolution

condition. However, the power of Koto ward declined once again and the governmental financial capacity was undercut in the fourth period.

The existence of a powerful claimant, Koto ward, was a major reason why the idea of siting incinerators in every ward became influential in the first and third periods. Koto ward was capable of influencing the policy-making process in waste management by making use of the de facto veto power over the new landfills and the threat of the blockade of waste. Because the delay in the new landfills' siting and the blockade of waste into the existing ones would have led to the collapse of the whole waste disposal process, the strategy of taking the landfills as a 'bargaining chip' was effective in the negotiation with the TMG. In other words, it was the garbage crisis in the early 1970s and the 1990s that provided the ward with the opportunity to make its claim heard in the policy-making process. Conversely, once the issue on the new landfills' siting was settled and the sense of emergency was fading away in the second and fourth periods, the power of Koto ward became weaker.

The power of the union was a significant factor, especially for the idea of devolving the responsibility to each ward. In the first and second periods, it was the union which obstructed the idea of the self-responsibility of each ward for waste disposal from influencing the policies. The union was able to make its claim heard by the government, because, as a socialist local government, the Minobe administration could not ignore a claim from one of its most important supporters. On the other hand, in the third period, the union persistently demanded that IWWD, both siting incinerators in every ward and the institutional responsibility of each ward, be implemented as a precondition for the devolution in order to stop this institutional change. The union was powerful enough to impose IWWD on the autonomy reform despite the reluctance of the TMG and the 23 wards, for it was granted political veto power by the Ministry of Home Affairs in the policy-making process of the entire autonomy reform. However, when the union started explicitly protesting IWWD once again to stop the devolution as OWOI was compromised in the 1997 revision, the 23 wards managed to overcome the union's opposition in 1998, which might have worked positively for the idea of the self-responsibility of each ward.

The power of the governments (i.e. the TMG, and the 23 wards after the devolution in 2000) was crucial for the dominance of IWWD as well. Even though IWWD was adopted in the siting policies in the first and third periods, it had to be compromised in the second and fourth periods because the governments were not able to implement it as planned. In the second period, the inability of the TMG to quickly resolve the local resistances to the incinerator projects led to the revision of the siting plan in 1974. The scarce land availability and financial difficulty was detrimental to the governmental capacity as well. Finding sites for the projects was difficult, especially in the hyper-congested central area of Tokyo. The financial crises in the second and fourth periods undermined the governmental capacity to afford as many incinerator projects as IWWD required. Thus, while IWWD was accepted by the governments, their limited ability to put this idea into practice undermined the strength of this idea of distributive justice in the second and fourth periods.

Exogenous Environments

The environments outside of the political system of Tokyo had significant impacts on the dominance of IWWD as shown in Table 7.5. The changes in the economy and the national policy contributed to changes in the dominance of the idea of OWOI. In the first period, the rapid economic growth and incinerationism as the policy paradigm worked for the rise of the idea of siting incinerators in every ward. In the second period, the slowdown of the economic growth inflicted negative impacts on the dominance of the idea. On the other hand, the national politics in the local autonomy reform worked negatively for the institutional responsibility of each ward in the first and second periods. In the third period, the economic boom helped the resurrection of IWWD, while the rise of Sustainable Waste Management (SWM) gradually undermined the dominance of the idea. Furthermore, the resistance of MOH to the local autonomy system of Tokyo made a favourable political environment for both siting incinerators in every ward and the self-responsibility of each ward. In the fourth period, however, the combination of the prolonged economic recession, the further advancement of SWM, and the dioxin crisis constituted the disadvantageous environments for IWWD as a whole.

For the dominance of the idea of siting incinerators in every ward, the most significant environmental factors were the economic conditions and the waste management policies at a national level. The strength of OWOI fluctuated along with the economic changes. When the economy boomed, the influence of IWWD in the siting policies increased, and vice versa. In the first period, IWWD emerged and rose to prominence under the rapid economic growth which Japan had enjoyed since the 1950s. In the second period, however, its policy influence fell along with the shift to reduced economic growth resulting from the two oil crises. Although the booming bubble economy in the latter half of the 1980s resurrected the influence of IWWD in the third period, it started falling again during the economic recession after the bubble burst in the fourth period. The rise and fall of the influence of IWWD synchronised with these economic fluctuations, although there were certain time lags.

The national policy in waste management constituted another significant environment for IWWD. As the Japanese political system was highly centralised, policies at the local level were strongly influenced by national policies. With regard to the dominance of IWWD in Tokyo, the shift in the policy paradigm was significant. In Japan, incinerationism had been dominant as the policy paradigm in waste management. However, SWM became influential as a new paradigm in the 1990s.

Incinerationism provided a favourable ideational environment for the idea of siting incinerators in every ward. This paradigm was characterised by the technological belief in incineration. The Japanese government had strongly encouraged local municipalities to construct incinerators by providing subsidies since the early 1960s. It was under this national policy to promote incineration that the TMG set AWI as the primary policy goal in waste management. IWWD was always associated with the need for more incinerators to achieve AWI; the dominance of

Table 7.5 Exogenous environments and dominance of IWWD

	First period (1971–1973)	Second period (1974–1989)	Third period (1990–1996)	Fourth period (1997–2003)
• For siting incinerators in every ward	+ • Rapid economic growth (+) • Incinerationism (+)	− • Economic slowdown (−) • Incinerationism (+)	+ • Economic boom (+) • Rise of SWM (−) • Attitudes of MOH (+)	− • Long economic recession (−) • Rise of SWM (−) • Dioxin disturbance (−)
• For institutional responsibility of each ward	− • Influence of JSP (−)	− • Influence of JSP (−)	+ • Attitudes of MOH (+)	− • Attitudes of MOH (+) • Dioxin disturbance (−)

JSP: Japanese Socialist Party; MOH: Ministry of Home Affairs

IWWD was tightly connected with this policy paradigm, which necessitated more incinerators amidst the rapid increase of waste.

Accordingly, the change in the policy paradigm affected the dominance of IWWD. In the third period, Japan experienced a paradigm shift in waste management; SWM was emerging as the new policy paradigm. This paradigmatic policy idea was to extend the scope of waste management to the process of waste production, and it involved up-stream measures such as reduction, reuse, and recycling (3Rs), in contrast to incinerationism, which was only concerned with the processes after waste was produced. This rise of SWM worked negatively for IWWD, which was tightly connected with incinerationism, as it undermined the necessity for more incinerators by reducing not only the actual amount of waste, but also lowering the estimated amount in the future. Although the impact of this nation-wide shift in the policy paradigm was still insufficient to set off the impact of the economic boom in the third period, it was gradually undercutting the very basis of the idea of siting incinerators in every ward.

In addition, in the fourth period, the dioxin crisis and the resultant policy change at the national level had negative impacts on both the dominance of siting incinerators in every ward and the institutional responsibility of each ward. Firstly, as small incinerators were regarded as the major source of dioxin pollution, the Ministry of Welfare encouraged a small number of large incinerators, rather than a large number of small incinerators. This policy change undercut the chance for the 23 wards to obtain the national subsidy for the new incinerators. Secondly, some existing incinerators were renovated to satisfy the tightened regulation for dioxin prevention issued by the Ministry of Welfare. This reduced the incineration capacity in Tokyo temporarily and led to the postponement of the shift to the self-sufficient disposal system outlined in the 1998 agreement.

For the institutional responsibility of each ward, the national politics on the local autonomy reform constituted the most significant environment outside of Tokyo, because devolving the authority of waste management required the amendment of the Local Autonomy Act. Through the first and second periods, the exogenous political environment worked negatively for the idea of the institutional responsibility of each ward to become influential. Although the Local Autonomy Act was amended in 1964 and 1974, the devolution of the authority in waste management was not realised as the Japanese Socialist Party won the supplementary provision, which left waste related services in the hands of the TMG.

In the third and fourth periods, the attitude of the MOH towards the local system reform of Tokyo made a significant political environment for IWWD. To achieve the local system reform, the TMG and the 23 wards had to persuade the MOH, who had the jurisdiction over the Local Autonomy Act. The ministry imposed two conditions: the devolution of waste management to each ward and the agreement of the concerned parties over the devolution. These conditions provided a favourable political environment in which IWWD became dominant in the policies of waste management and devolution.

Interaction between the four variables

As detailed in the previous chapters, the empirical results demonstrate that the four explanatory variables interacted with one another in the policy-making processes, and their interaction had significant impacts on the influence of IWWD. One of the most notable features of the result is that all of the four explanatory variables went up and down in synchronisation for the idea of siting incinerators in every ward, which led to the clear rise and fall of IWWD. The interaction between the variables centred on the economic conditions explains why the four variables changed in synchronisation.

The economic situation affected the ideational legitimacy of IWWD by influencing the production of waste. IWWD was recognised as cognitively legitimate only when the amount of waste was rapidly increasing and the growth was expected to continue in the future. Accordingly, when the growth of waste slowed down, or the growth was not expected in the future, the ideational legitimacy of IWWD became weak. As the production of waste was closely related to economic conditions, the degree of the ideational legitimacy of IWWD went up and down along with the economic fluctuations.

Furthermore, the degree of the interest of Koto ward was significantly affected by the amount of waste production. The rapid growth in the quantity of waste was one of the causes that made Koto ward advocate IWWD. The increasing amount of waste under the booming economy resulted in the influx of un-incinerated waste into the landfills next to Koto, causing environmental pollution. When the increase of waste slowed down and AWI was expected to be achieved without constructing as many incinerators as IWWD projected, the interests of the ward in this notion of distributive justice also declined.

The power of Koto ward and the governments was also partly dependent on economic situations. The economic prosperity stimulated waste generation and produced the urgent necessity for the new landfills, over which Koto had de facto veto power in the first and third periods. When the crises abated in the second and the fourth periods, the power of the ward in the negotiation with the TMG declined. The economic situation also affected the governmental capacity to realise IWWD. While the financial capacity of the governments was robust under the economic booms, the economic downturns damaged their tax revenue and made the governments unable to proceed with the OWOI policy in the second and fourth periods.

Besides the interaction centred on the economic conditions, the four variables interacted in other ways too. For instance, the cognitive legitimacy of the idea of siting incinerators in every ward was influenced not only by the economic conditions and the resultant production of waste, but also by the policy paradigms in waste management. The ideational legitimacy of the idea of siting an incinerator in every ward was closely related to incinerationism, the long-held policy paradigm in waste disposal. Accordingly, the rise of SWM as a new policy paradigm undermined the cognitive legitimacy of IWWD.

Another example is the dioxin problem and resultant national policy change affecting the cognitive legitimacy of the self-responsibility of each ward and the financial power of the 23 wards. In the fourth period, the dioxin problem led to a temporary reduction in incineration capacity, which made it impractical to shift to even the block incineration system as planned in the 1994 agreement; the dioxin crisis weakened the cognitive legitimacy of devolving the institutional responsibility to each ward. The dioxin problem also caused the national policy shift towards regional disposal, which affected the financial capability of the 23 wards to construct incinerators, due to a reduced chance of obtaining national subsidy for new incinerators.

Furthermore, in the third period, the interest of the union interacted with the other variables and resulted in the culmination of IWWD in its influence on policies. Contrary to the first and second periods, the interest of the union worked positively for IWWD in the third period. The union imposed IWWD as a precondition for the devolution from its strategic intention to make this institutional reform impossible to attain. The union adopted this strategy because it was granted a political veto in the process of local autonomy reform by the MOH and knew that IWWD was difficult to implement (weak cognitive legitimacy), but at the same time normatively irrefutable (strong normative legitimacy). What engendered the political driving force for IWWD in the third period was this interaction between the powerful interest of the union, the ambivalence in the ideational legitimacy of the idea, and the politics over local autonomy, both inside and outside of Tokyo.

Discussion and implications

Power conflicts of interests or ideational causes?

Pluralists and rationalists argue that interests and power of carriers are more decisive for the dominance of an idea, while scholars in ideas and politics emphasise the importance of the characteristics of an idea for its prominence, since they mobilise potential adherents and demobilise antagonists. The empirical result demonstrates that it is necessary to grasp both the power struggles between interests and the characteristics of an idea.

This empirical study confirms the importance of a powerful claimant who has strong interests in an idea and is able to influence the policy-making process. Compared with policy paradigms and worldviews, IWWD is a policy idea which is high-profile at the low level of generality, open to everyday policy debates, and therefore more subject to calculating, strategic behaviours for achieving self-interested goals (Campbell, 1998, 2002; Mehta, 2010; Schmidt, 2008). It was the claims and campaigns of Koto ward for IWWD that made the policy makers take this idea seriously. IWWD would not have been adopted into policies without the strong interest and power of Koto ward. The changing dominance of the idea of the self-responsibility of each ward also demonstrates the significance of powerful interests. The idea of the self-responsibility of each ward rose to prominence in the third period due to the powerful interest of the union, although all of the actors, including the union, thought that waste disposal was better performed regionally rather than locally.

Then, are ideas mere "hooks" for interests? Do the characteristics of the ideas matter at all? The empirical result of IWWD also illustrates the importance of the characteristics of the ideas. OWOI became influential in the policies not only because Koto had strong interest and power, but also because this idea of siting an incinerator in every ward was ideationally appealing to the policy makers. If IWWD had not fitted with the values which underlay the siting scheme of the TMG and/or had not been recognised as a good idea to solve the garbage problems, IWWD would not have been adopted that quickly and influentially. The cognitive legitimacy was especially crucial for the policy makers who were in charge of urban management. It was this synergy between the power and interest of Koto ward and the ideational legitimacy of IWWD that led to this idea being quickly adopted into the policies.

Similarly, the rise of the idea of the self-responsibility of each ward in the third period cannot be attributed to the interest and power of the union alone. It is true that the power and interest of the union made IWWD dominant in policies, despite the weak cognitive legitimacy of the idea. However, the self-responsibility of each ward strongly resonated with the notion of the autonomy of the wards, which the 23 wards had long pursued. Although devolving the responsibility in waste disposal was not regarded as a good idea in the cognitive sense, this idea was normatively compelling for the 23 wards amidst the height of the autonomy expansion movement. Furthermore, as noted above, the interest of the union worked positively for IWWD in the third period as it interacted with the cognitively weak but normatively strong ideational legitimacy of IWWD and the national political environment of local autonomy reform.

Thus, even though IWWD is a policy idea which is more likely to be subject to interests and power, the dominance of this idea cannot be explained without considering the role of the normative/cognitive characteristics of the idea. Politics is both powering and puzzling. To better understand the politics of ideas, it is necessary to grasp how both the power struggles between interests and the contents of an idea affect the chances of its success in policies.

Economic determinism?

As noted earlier, the economic factor played a central role in the interaction between the variables. The economic booms caused IWWD to be perceived as cognitively legitimate, strengthened its congruence with the interest of Koto ward, empowered this claimant in the policy-making process, and improved the financial capacity of the governments to implement the incinerator projects. Accordingly, the economic downturns affected these variables in the opposite way, and thus undermined the dominance of IWWD in the policies. The economic shifts account for a substantial portion of the changes in the dominance of IWWD, triggering the changes in the other variables. Then, does this mean that the economy determines the dominance of an idea? This question requires a careful consideration.

Firstly, the influence of policy paradigms must be considered. The resonance with ideas at higher levels of generality affects the chance for a policy idea at the lower levels of generality, such as IWWD, to become influential in policies. The ideational legitimacy of the idea of siting an incinerator in every ward was significantly dependent on incinerationism, the dominant policy paradigm in waste management. It was through this interpretive framework that the soaring amount of waste under the economic booms was translated into the urgent necessity for more incinerators and that OWOI was perceived as a good idea to solve the waste problems. In contrast, SWM was to decouple the economic growth and the production of waste, although the development of the 3Rs under this new policy paradigm was not enough to set off the impact of the economic boom in the third period. SWM provided alternatives to expanding incineration capacity to tackle the growth of waste. Thus, it depended on the policy paradigms to determine whether the economic growth led to the robust cognitive legitimacy of the notion of siting incinerators in every ward. It was the interaction between the material and ideational factors which determined the cognitive legitimacy of siting incinerators in every ward.

Secondly, for the self-responsibility of each ward, the changes were caused by the political factors around the devolution and the local autonomy reform of the 23 wards. In the first and second periods, the interest and power of the union played a decisive role in preventing this notion of IWWD from influencing the policies. In the third period, to the contrary, the same interest and power boosted the dominance of the self-responsibility of each ward. Interacting with cognitively weak but normatively strong ideational legitimacy, the interest of the union resulted in the strategy of imposing IWWD as a precondition for the devolution to stop this institutional reform. This strategy worked as the union was granted political veto in the political process of the autonomy reform. Moreover, this rise of the self-responsibility of each ward led to the further acceleration of the OWOI policy. The influence of IWWD reached its culmination in this period not only because of the economic booms, but also because of this political development of the autonomy reform.

Towards further study on politics of ideas and siting conflicts

Through elucidating the causes and mechanisms behind the changing fortunes of IWWD in the 23 wards of Tokyo, this study paves the way for further study on politics of ideas and siting conflicts in several respects.

Firstly, the empirical result shows the importance of taking into account different types of variables and examining the interaction between them to explain the dominance of an idea and its change over time. The changes in the dominance of IWWD cannot be explained by a single type of variable. While the interests and power of the actors were indispensable for IWWD to rise to prominence, normative/cognitive ideational legitimacy played a significant part as well. Without knowing how the economic changes affected the other variables, the rise and fall of the dominance of IWWD cannot be understood. On the other hand, the economic factors alone cannot explain the whole picture, either. The cognitive legitimacy of the idea was significantly influenced by policy paradigms as well. The changes in the dominance of the self-responsibility of each ward were caused by the interests and power of the union interacting with the ambivalent ideational legitimacy and the political environments around the local autonomy reform of the 23 wards. The dominance of IWWD rose and fell as a result of the complicated interaction between the characteristics of the idea, the interests and power of the actors, and the economic, political and ideational environments. While ideational approaches in political studies have focused on proving causal influence of ideas in political phenomenon and outcomes (Béland & Cox, 2010), we need to incorporate other types of factors in the analysis and examine the interaction between them to better understand politics of ideas.

Secondly, this research indicates the importance of exploring the policy-making and implementation processes to further environmental justice studies. Environmental justice studies have focused on proving the existence of environmental inequality and exploring causes and mechanisms of the spatial and racial injustice. For instance, Pellow (2000, 2004) shows that workers in a recycling facility of Chicago had suffered poor and dangerous working conditions as a result of a large-scale recycling programme collaboratively supported by environmentalists, the state, community groups, and industrialists, and thereby proposes the Environmental Inequality Formation perspective to go beyond perpetrator-victim scenario. On the other hand, how and why a particular governmental policy to address environmental injustice is formed has been less studied in the literature. Furthermore, as the case of IWWD showed, a policy to rectify distributive inequality often provokes sharp conflicts because it changes the distribution of costs and benefits among communities and groups; the institutionalisation of an idea of distributive justice is not the end of the story, but the introduction to further conflicts. To fully understand the politics of environmental justice and develop better policy solutions to distributive inequality, it is essential to explore not only the formation of inequality, but also the formation process of a distributive justice policy and the difficulties that policy faces through its implementation process.

Thirdly, we need to understand siting conflicts and policies for distributive justice in a more contextualised way. As demonstrated through the empirical study, the success of an idea of distributive justice in siting policies depends on various intertwining factors such as economic situation, policy paradigm, political institution, and interests and power of stakeholders. The economic situation affected the validity of IWWD through impacting waste production and governmental financial conditions. There was the mechanism through which the economic changes led to the change in the interest and power of Koto ward. IWWD was closely connected with incinerationism as the long-held policy paradigm in Japan. This idea of distributive justice was deeply embedded into the political system of the 23 wards, which had developed an autonomy expansion movement intended to change their limited status as local municipalities. Distributive justice is a significant issue in siting conflicts and various ideas and approaches have been proposed by scholars and practitioners (Been, 1992, 1994; Kunreuther, 1986; Morell, 1984; Portney, 1991; Rabe 1994). However, an idea of distributive justice which succeeds in a place and time may nonetheless fail in another place and time; it is necessary to carefully consider economic, policy and political contexts of a concerned society to find a better solution to distributive issues in facility siting.

Last, but not least, this research lays a foundation for further comparative research. This study is just one case study on one particular idea and place; the result cannot simply be generalised. As the dominance of IWWD was closely related to the conditions and mechanisms particular to the 23 wards of Tokyo, the empirical arguments of the study are not necessarily directly applicable to other cases. Nonetheless, the comprehensive framework and in-depth empirical analysis of this study paves the way for further comparative research with other municipalities in Japan and of other countries. Conflicts over locally unwanted facilities are a significant problem and distributive justice is a crucial aspect in them. Siting of waste disposal facilities is a pressing issue, especially in fast-growing Asian countries that are facing drastic increase of waste and rapid urbanisation. Principles similar to IWWD have been proposed and adopted in policies to deal with distributive issues in many places throughout the world. New York City adopted the fair share approach (Valletta, 1993). In the European Union, the proximity principle and the self-sufficient principle have been recognised as significant concepts in waste management (Watson and Bulkeley, 2005). In Seoul, an idea that incinerators should be sited in every ward was declared in 1990s (Moon et al., 2006). The jurisdictional responsibility was introduced to manage conflicts over waste disposal facilities in Beijing (Běijīng Shì Rénmín Zhèngfǔ, 2013). The book, by elucidating how IWWD's influence in Tokyo waxed and waned along with the economic, policy and political changes, provides comparative perspectives to understand, and find a solution to, siting conflicts and waste management problems.

References

Been, V. (1992). What's Fairness Got to Do with It? Environmental Justice and the Siting of Locally Undesirable Land Uses. *Cornell Law Review*, 78, 1001–1085.

Been, V. (1994). Conceptions of Fairness in Proposals for Facility Siting. *Maryland Journal of Contemporary Legal Issues*, 5, 13–24.

Běijīng Shì Rénmín Zhèngfǔ. (2013). Běijīng Shì Shēnghuó Lèsè Chǔlǐ Shèshī Jiànshè Sān Nián Shíshī Fāng'àn (2013–2015 nián).

Béland, D., & Cox, R. H. (2010). Introduction: Ideas and Politics. In D. Béland & R. H. Cox (eds.), *Ideas and Politics in Social Science Research*. New York: Oxford University Press.

Campbell, J. L. (1998). Institutional Analysis and the Role of Ideas in Political Economy. *Theory and Society*, 27(3), 377–409.

Campbell, J. L. (2002). Ideas, Politics, and Public Policy. *Annual Review of Sociology*, 28(1), 21–38.

Kunreuther, H. (1986). A Sealed-bid Auction Mechanism for Siting Noxious Facilities. *The American Economic Review*, 76(2), 295–299.

Mehta, J. (2010). The Varied Roles of Ideas in Politics from "Whether" to "How." In D. Béland & R. H. Cox (eds.), *Ideas and Politics in Social Science Research*. New York: Oxford University Press.

Moon, D., Shirakawa, H., Tabata, T., & Imura, H. (2006). Study on the Information Needs of Residents in EIA of the Incineration Facility Construction in Seoul City. *Proceedings of the 20th Conference on Environmental Information Science*, 20, 415–420.

Morell, D. (1984). Siting and the Politics of Equity. *Hazardous Waste*, 1(4), 555–571.

Pellow, D. N. (2000). Environmental Inequality Formation: Toward a Theory of Environmental Injustice. *American Behavioral Scientist*, 43(4), 581–601.

Pellow, D. N. (2004). *Garbage Wars: The Struggle for Environmental Justice in Chicago*. Cambridge: The MIT Press.

Portney, K. E. (1991). *Siting Hazardous Waste Treatment Facilities: The Nimby Syndrome*. New York: Auburn House.

Rabe, B. G. (1994). *Beyond Nimby: Hazardous Waste Siting in Canada and the United States*. Washington D.C.: Brookings Institution Press.

Schmidt, V. A. (2008). Discursive Institutionalism: The Explanatory Power of Ideas and Discourse. *Annual Review of Political Science*, 11(1), 303–326.

Valletta, W. (1993). Siting Public Facilities on a Fair Share Basis in New York City. *The Urban Lawyer*, 25(1), 1–20.

Watson, M., & Bulkeley, H. (2005). Just Waste? Municipal Waste Management and the Politics of Environmental Justice. *Local Environment*, 10(4), 411–426.

Index